Volker Schulz
Advanced Hydrogel-based Chemical Microsensors

TUD*press*

DRESDNER BEITRÄGE ZUR SENSORIK

Herausgegeben von
Gerald Gerlach

Band 50

Volker Schulz

Advanced Hydrogel-based Chemical Microsensors

TUD*press*
2013

Die vorliegende Arbeit wurde am 31. Januar 2013 als Dissertation an der Fakultät Elektrotechnik und Informationstechnik der Technischen Universität Dresden eingereicht und am 10. Juli 2013 erfolgreich verteidigt.

Gutachter:

Prof. Dr. phil. nat. habil. R. Tetzlaff (Vorsitzender)
Prof. Dr.-Ing. habil. G. Gerlach (Gutachter)
Prof. Dr.-Ing. habil. T. Wallmersperger (Gutachter)
Prof. Dr.-Ing. A. Richter

Bibliografische Information der Deutschen Nationalbibliothek
Die Deutsche Nationalbibliothek verzeichnet diese Publikation in der Deutschen Nationalbibliografie; detaillierte bibliografische Daten sind im Internet über http://dnb.d-nb.de abrufbar.

Bibliographic information published by the Deutsche Nationalbibliothek
The Deutsche Nationalbibliothek lists this publication in the Deutsche Nationalbibliografie; detailed bibliographic data are available in the Internet at http://dnb.d-nb.de.

ISBN 978-3-944331-30-0

Verlag der Wissenschaften GmbH
Bergstr. 70 | D-01069 Dresden
Tel.: 0351/47 96 97 20 | Fax: 0351/47 96 08 19
http://www.tudpress.de

Gesetzt vom Autor.
Printed in Germany.

Editor's Preface

Hydrogels are polymer materials which can take up large amounts of a solvent and concurrently maintain their dimensional stability, i.e. they do not dissolve in the solvent. Especially for stimuli-sensitive hydrogels, the amount of absorbed solvent can be triggered by a variety of external parameters, e. g. temperature or the concentration of different analyte species in aqueous solutions. Because of this property, hydrogels are very interesting materials for the application in sensors and actuators where the stimulus-specific swelling is deliberately used. Hydrogel-based sensors are simple devices which are robust and cost-efficient. However, their specific structure and the accompanied properties result in particular challenges which have to be solved to apply these sensors in real-world applications. These properties are e.g. the cooperative movement of the polymer chains in the solvent which leads to long sensor response times as well as viscoelastic chain relaxation which affects the sensor's long-term stability.

The author of this work faces up to these challenges and presents different approaches to overcome the problems stated, such that hydrogel-based sensors might already find market access in the near future. In this sense, this work is not only very interesting from a scientific point of view but also relevant from a technical perspective. Therefore, I truly wish that this work receives the attention it deserves.

Dresden, July 2013 Gerald Gerlach

Author's Preface

The current work is the result of a research project conducted during my time as a research associate at the Solid-State Electronics Laboratory (IFE), Technische Universität Dresden and was financially supported by the German Research Foundation (DFG) under grant GE779/24-1. This work goes along with research work performed on hydrogel-based sensors at the IFE in the last decade and builds up on the profound knowledge developed there. Here, the aim is the improvement of the sensor properties of hydrogel-based sensors where different technological solutions and their distinct implementation are proposed. An improved approach to this interdisciplinary subject is not possible without a close and vivid cooperation between chemists, material scientists and engineers and would not have been successfully established without the competence, support, encouragement and unremitting motivation of many colleagues, students and friends. Thank you all!

I wish to express my particular thanks and appreciation to:

- my Ph.D. supervisor Prof. Dr.-Ing. habil. Gerald Gerlach for the possibility of writing my thesis at the laboratory, the excellent scientific support, the scientific freedom where many ideas had the possibility of growing, and the extraordinarily pleasant cooperation on a personal level,
- Prof. Dr.-Ing. habil. Thomas Wallmersperger (Professur für Mechanik multifunktionaler Strukturen, TU Dresden) for fruitful discussions and for kindly agreeing to be the second reviewer,
- Dr. Stefan Zschoche (Leibniz-Institut für Polymerforschung Dresden e. V.) for his enduring motivation and his enthusiasm in preparing macroporous hydrogels as well as for inspiring discussions,
- Dr. Bernhard Ferse and Dipl.-Chem. Anna Große (Professur für spezielle physikalische Chemie – Physikalische Chemie der Polymere, TU Dresden) for their time, motivation and skills in voluntarily preparing nanocomposite hydrogels beside their own ambitious projects and everyday routines,
- Prof. Dr.-Ing. habil. Dipl.-Math. Klaus Röbenack (Institut für Regelungs- und Steuerungstheorie, TU Dresden) for a pleasant cooperation and useful discussions on system's theory and force compensation,
- Dr. Jörg Sorber who has always been a source of inspiration an ideas to me

and has become more than a colleague over the years,

- all colleagues of IFE for their kind support in theoretical and practical issues, especially PD Dr.-Ing. Margarita Günther, Dipl.-Ing. Volker Krause, Dipl.-Ing. Siegfried Kostka, Ulrike Lehmann, and Sabine Herbst,
- all former students who directly and indirectly enriched this work with their thesis results or practical lab-work (Dipl.-Ing. André Juhrig, Dipl.-Ing. Henning Ebert, Dipl.-Ing. Qiang Ren, Dipl.-Chem. Haiping Zhang, Dipl.-Chem. Tom Teichert, M.Sc. Marion Schneider),
- Subramanian Ramanathan (now with the Department of Biosystems Science and Engineering, ETH Zürich) and Dr. Jens Friedrichs (Leibniz-Institut für Polymerforschung Dresden e. V.) for AFM-Nanoindentation measurements, Marion Stohr (Institut für Textilmaschinen und textile Hochleistungstechnik) for uniaxial tensile measurements, Dr. Martin Waegner (IFE) for REM measurements, Andreas Janke (Leibniz-Institut für Polymerforschung Dresden e. V.) for AFM analysis, M.Sc. Matthias Neumann (Professur für technische Chemie) for mercury intrusion porosimetry measurements,
- Anja, Martin, Marco, Andi, Yvonne, Christian, Micha, Adi, Markus, Sebastian, Mario, Anna, Alex, Jona, Oleg for a truly amazing and unforgettable time at the lab,
- the German Research Foundation (DFG) for funding this work,
- my parents, Martin, Anja, Julius and Petra for their everlasting support, their understanding, and their love.

Dresden, July 2013 Volker Schulz

Abstract

This work concerns hydrogel-based piezoresistive chemical microsensors which particularly employ stimuli-sensitive hydrogels as functional sensing layers for the detection of most diverse analytes (e. g. pH) in aqueous solutions. The implementation of these sensors in real world applications is still challenging due to the existence of long sensor response times, limited mechanical stability of the hydrogel sensing layers, and dynamic long-term drift effects. In this work, two different approaches for the improvement of the sensor properties are addressed. A material-based approach aims at the specific improvement of the hydrogel material properties. In particular, a macroporous, interconnected network structure with pore diameters in the µm-range is incorporated into a poly(acrylamide-*co*-acrylic acid) hydrogel by the porogen-leaching method in order to improve the sensor response time. With this method, a response time reduction of about 60 % could be achieved. Furthermore, the mechanical stability of poly(*N*-isopropylacrylamide) sensing layers is greatly improved by the formation of physically cross-linked, hybrid networks using inorganic Laponite XLS clay nanoparticles as multifunctional crosslinks. It is shown that these sensors exhibit an improved response time for temperature stimulation and measurements in organic co-solvents like ethanol. A second approach applies the force compensation method to hydrogel-based sensors. Using a system-based approach, the influence of a closed-loop feedback structure on important figures of merit in sensor technology, like sensitivity, response time, suppression of dynamic long-term drift effects and systematic static deviations, is described on a theoretical basis. The resulting conclusions are used for the design of a force-compensated hydrogel sensor in closed-loop configuration. In this approach, the pH-induced osmotic swelling pressure of a poly(vinyl alcohol)/poly(acrylic acid) hydrogel is counteracted by an appropriate compensation pressure during the whole measuring process such that the hydrogel volume is kept constant in the steady state. In this way, the macroscopic cooperative movement of the polymer chains in the measuring solution is suppressed and, as a consequence, the sensor response time and long-term drift effects are significantly reduced.

Contents

List of symbols

a	activity, correction term, geometric dimension, coefficient
b	geometric dimension, coefficient
c	concentration, coefficient
d	dimension, thickness, diameter, deflection
d_c	critical dimension, coefficient
d_pore	pore diameter
e_c	controller input signal
f	friction coefficient, non-linear sensor part
f_s	shape factor
f_Q	quantization error
f'_Q	reduced quantization error
g_m	gain margin
g_s	impulse response of the sensor
h	thickness
k	BOLTZMANN constant, spring constant, rate constant
k_P	P-factor
k_I	I-factor
k_D	D-factor
n	number of moles, number of bits, compliance
p	pressure
r	radius, relative resistance change, radial coordinate, real part of dominant pole
$\boldsymbol{r}$	position vector
$\langle r \rangle$	time-averaged contact radius
s	number of segments of polymer chains, complex frequency
s_p	dominant pole
t	time
t_exp	exposure time
t_s	settling time
t_{90}	time where 90 % of the maximum value is reached
$\boldsymbol{u}$	displacement vector
w	deflection, geometrical dimension
Δw	displacement

x	one-dimensional spatial coordinate, signal
Δx	compensation deviation
x_{A}	general actuator input signal
x_{off}	static sensor offset
$\tilde{x}_{\mathrm{off}}$	static sensor offset subjected to deviations
y	spatial coordinate, output signal
z	spatial coordinate, charge number
A	mobile anions, amplification factor
ΔA	drop in signal amplitude
A_{p}	linear stiffness coefficient
B_{p}	non-linear stiffness coefficient
C	mobile cations
D	diffusion coefficient, flexural rigidity
$D(s)$	denominator polynomial
D_{coop}	cooperative diffusion coefficient
D_{bm}	diffusivity of the buffer medium
E	YOUNG's modulus
F	force, quantity which correlates the input signals of a closed-loop sensor
F_{counter}	counter force
F_{gravity}	gravity force
F_{L}	loading force
FB	feedback factor
G	GIBBS free energy (free enthalpy), shear modulus, transfer function
G^{Δ}, $G^{\Delta X}_{X_{\mathrm{i}}}$	transfer function between the compensation deviation and the sensor input signal
$G'_{\mathrm{s}}(s)$	generalized sensor transfer function
$\tilde{G}_{\mathrm{s}}$	open-loop sensor transfer function of a closed-loop sensor, real transfer function of the hydrogel
$G^{\Delta\mathrm{K,d}}(s)$	transfer function considering a time-dependent sensitivity due to drift
H	enthalpy
I	ionic strength
$\mathrm{Im}\{s\}$	imaginary part of s
J	performance index
J_{conv}	convective flow
K	bulk modulus, K-factor, static sensitivity

$\tilde{K}$	static sensitivity subjected to deviations
K^*	amplified static sensitivity
K_a	acid dissociation constant
$\mathcal{L}$	LAPLACE operator
M_n	molecular weight
N	number of charge carriers
$N(s)$	numerator polynomial
N_A	AVOGADRO constant
N_c	number of polymer chains
Q	quality factor, swelling degree
R	universal gas constant, ratio of shear and bulk modulus, resistance, radius
$\Re\{s\}$, $\mathrm{Re}\{s\}$	real part of s
S	entropy, sensitivity
T	temperature, time constant
T_c	phase transition temperature
T_d	lag time
T_D	differential time, denominator time constant
T_I	integral time
T_N	numerator time constant
T_1, T_2, T_3	characteristic sensor time constants
V	volume, voltage, amplification factor
V_B	bridge voltage
V_c	control signal
$V_\mathrm{I,ADC}$	input voltage of the analog digital converter
V_off	offset voltage
V_s	supply voltage
X	signal in the frequency domain
ΔX	compensation deviation
X_c	compensation quantity
X_i'	generalized sensor input signal
$\Delta X_\mathrm{o}^{\Delta\mathrm{K,s}}$	deviation of the sensor output signal induced by static deviations of sensitivity
$\Delta X_\mathrm{o}^{\Delta\mathrm{K,d}}$	deviation of the sensor output signal induced by dynamic deviations of sensitivity
X_ref	reference quantity
Z	interference quantity
α	degree of ionization

α_o	offset coefficient
α_s	sensitivity coefficient
β	factor accounting for the influence of feedback deviations
β_bm	buffer capacity of the buffer medium
β_gel	buffer capacity of the hydrogel
γ	surface tension
δ	drift rate, indentation depth
$\delta G_\mathrm{u}(s)$	uncertainty of the hydrogel transfer function
ε	mechanical strain, number between 1.9 and 2.4
ε_d	systematic dynamic deviation
ε_s	systematic static deviation
η	viscosity
λ	isotropic expansion factor, wavelength
μ	chemical potential, charge carrier mobility
ν	POISSON's ratio, slope
ν	stoichiometric coefficient
ν_c	cross-linking density
π	piezoresistive coefficient
σ	mechanical stress
σ_{xx}, σ_{yy}	bidirectional swelling stress
τ	time constant
τ_bmd	time constant due to buffer mediated diffusion reaction
φ	volume fraction, ratio of sensor time constants
φ_m	phase margin
χ	FLORY-HUGGINS interaction parameter
Δ	(systematic) deviation, difference
Θ	contact angle
$\mathbf{\Pi}$	second order tensor containing piezoresistive coefficients

Indices

A	actuator
c	controller
comp	compensation
COO-	carboxylate groups
COOH	carboxyl groups
eff	effective

el	elastic contribution
eq	equilibrium
H+	hydrogen ions
i	ionizable groups, input
i	system component i
IL	intermediate bonding layer
ion	ionic contribution
j	system component j
l	longitudinal
m	molar, mixing contribution, mass
M	measuring unit
max	maximum value
o	output
out	output
p	polymer
ref	reference
s	solution, solvent, sensor
sp	sphere
swell	swelling
t	transversal
tot	total
v	volume
visc	viscous contribution
+	cations
-	anions
α	analyte species α

Superscripts

cl	closed-loop
comp	sensor operated with compensation method
defl	sensor operated with deflection method
ext	in the external solution
gel	inside the hydrogel
lin	linear
ol	open-loop
P	P-controller
PI	PI-controller
PID	PID-controller

List of abbreviations

AA	acrylic acid
AAm	acrylamide
AAO	anodized aluminium oxide
ADC	analog digital converter
AFM	atomic force microscope
ATR-FTIR	attenuated total reflection Fourier transform infrared spectroscopy
BIBO	bounded-input bounded-output
BIS	*N,N'*-methylenebisacrylamide
CCA	crystalline colloidal array
DMA, DMAEMA	*N,N*-dimethylaminoethyl methacrylate
DMAPAA	*N*-(3-dimethylaminopropyl)acrylamide
DMIAAm	2-(dimethyl maleinimido)acrylamide
DRIE	deep reactive ion etching
EDX	energy-dispersive X-ray
FEM	finite element method
FFT	fast Fourier transform
FRET	fluorescent energy transfer
GOx	glucose oxidase
HEMA	2-hydroxyethyl methacrylate
HPMA	hydroxypropyl methacrylate
IDEA	interdigitated electrode array
IOA	isooctyl acrylate
IR	infrared
ISE	integral of squared error
IUPAC	international union of pure and applied chemistry
KPS	potassium peroxodisulfate
LCST	lower critical solution temperature
LED	light emitting diode
LOH	left open half-plane
LSPR	localized surface plasmon resonance
LTI	linear time-invariant
MAA	methacrylic acid

MCT	mercury cadmium telluride
MEMS	micro electro-mechanical system
MIP	mercury intrusion porosimetry
MPS	3-methacryloxypropyltrimethoxysilane
NaPS	sodium peroxodisulfate
Nd-YAG	neodymium-doped yttrium aluminum garnet
NIPAAm	*N*-isopropylacrylamide
NMR	nuclear magnetic resonance
NRET	nonradiative energy transfer
OLED	organic light emitting diode
OSC	organic solar cell
P	proportional
PAA	poly(acrylic acid)
PAAm	polyacrylamide
PBS	phosphate buffered saline
PC	polycarbonate, personal computer
PCCA	polymerized crystalline colloidal array
PDE	partial differential equation
PDMA	poly(*N*,*N*-dimethylaminoethyl methacrylate)
PDMS	polydimethylsiloxane
PE	polyether
PECVD	plasma enhanced chemical vapor deposition
PEG	poly(ethylene glycol)
PET	polyethylene terephthalate
pH	*pondus hydrogenii*
PHEMA	poly(2-hydroxyethyl methacrylate)
PI	proportional-integral
PID	proportional-integral-derivative
PMAA	poly(methacrylic acid)
PNIPAAm	poly(*N*-isopropylacrylamide)
ppb	parts per billion
PS	polystyrene
PTFE	polytetrafluoroethylene
PVA	poly(vinyl alcohol)
PVBC	poly(vinylbenzyl chloride)
P2VP	poly(2-vinylpyridine)
RET	rubber elasticity theory
SEM	scanning electron microscopy
SISO	single-input single-output

TCR	temperature coefficient of the specific resistance
TCS	temperature coefficient of the sensitivity
TCπ	temperature coefficient of the piezocoefficients
TEGDMA	tetraethyleneglycol dimethacrylate
TEM	transmission electron microscope
TEMED	*N,N,N',N'*-tetramethylethylenediamine
USB	universal serial bus
UV	ultraviolet
VPT	volume phase transition

I Introduction and theory

1 Introduction

1.1 Background and motivation

The market share of products containing micro electro-mechanical systems (MEMS) is continuously growing and predicted to grow further with an annual growth rate of about 9 % over the next five years. The market value of MEMS products in 2005 amounted to $ 8 billion, with a share of 40 % for sensor applications, but is expected to rapidly increase of up to $ 40 billion in 2015 and $ 200 billion in 2025 [1]. To live up to these expectations, strategies to solve the economic conflict triangle, i. e. quality enhancement, economy of time, and cost reduction, are essential. Since microsensors are an inherent part in almost all fields of application like automotive, medical and life sciences, industrial and laboratory process control and many more, innovative solutions and novel unconventional sensing approaches are highly desirable. In particular, for emerging fields of application like bioMEMS, medical devices, point-of-care diagnostics, and lab-on-chip platforms, which require low costs as well as high accuracy and reliability at the same time, these requirements are challenging. For this reason, simple operational principles have attracted increasing attention. Therefore, a trend, off from conventional hard materials, like glass, silicon, or metal, towards soft polymeric materials can be registered in current research. Recent advances in polymer synthesis enable the fabrication of polymers as functional sensor layers with custom-made specific properties. This might push them to center stage of cost-efficient, easy-processable polymer-based sensors and merge the fields of macromolecular chemistry and MEMS technology to an interdisciplinary field with high innovation potential. This trend also reflects in others than the sensor field, e. g. for organic light emitting diodes (OLEDs) or organic solar cells (OSCs) [2, 3].

Especially, the fascinating material class of polymer gels, which is rather new in microsystem technology, has attracted increasing attention in the last years. These materials are characterized by their ability to take up large amounts of a solvent. If the solvent is water, they are called hydrogels. Hydrogels which respond to environmental changes of either physical or chemical nature, like change in solvent concentration (e. g. pH), electric field, or light with a reversible volume or shape change with high sensitivity, are called "smart" or "intelligent" hydrogels due to their combined sensor-actuator functionality. These responsive gels have proved to be suitable candidates as functional sensor layers, especially for chemical and biological sensing in liquid environments. The possibility of a chemical design of these materials enable the fabrication of hydrogels which are sensitive to a large number of different analytes. Responsive hydrogels have the potential for miniaturization. Methods for the realization of defined hydrogel structures with arbitrary shape, from the millimeter to the sub-micrometer range, as well as methods for the fabrication of hydrogel thin films are already available [4]. MEMS can show integrated mechanical and electrical functionality on one chip. The combination of these outstanding properties facilitates the integration of hydrogels into microelectronic technology and enables the fabrication of hydrogel-based microsensors as a new generation of MEMS. The exploitation of synergies, i. e. the capability of mass production and miniaturization, the reliability of microelectronic technology, low costs, and specifically tunable polymer properties in a wide range, enables sensors suitable to replace existent bulky and costly solutions. In this way, biomedical devices for *in vivo* metabolic monitoring, affordable even for single patients, disposable sensors for biotechnological analysis or point of care diagnostics as well as in-line capable sensors for process monitoring can be established.

Especially, the combination of piezoresistive silicon pressure sensor chips with hydrogel sensing layers, as shown by GERLACH *et al.*, has proofed to be a promising approach towards the realization of robust, biocompatible, low-cost, in-line-capable sensors for monitoring of most different analytes in aqueous solutions [5, 6]. The basic sensor principle relies on the volume change of the hydrogel due to changes of certain quantities in the external measuring solution like pH, ionic strength or solvent concentration. The mechanical energy applied by the hydrogel is directly converted into an electrical output voltage by a piezoresistive pressure sensor chip which acts as mechano-electrical transducer. Current research results strongly suggest that the sensor properties are primarily determined by the physico-chemical properties of the hydrogel sensing layer. Besides intensive studies of the influence of hydrogel type, hydrogel film thickness, and film preparation on the sensor response time and sensitivity as well as hydrogel and sensor modeling [7–9], some key aspects, like long sensor response times due to cooperative diffusion of the polymer chains in the measuring solution, poor mechanical hydrogel stability – which means poor repeatability and long-term stability of the sensor signal – and dynamic drift effects due to the visco-hyperelastic material properties, are still challenging and hard to overcome. These challenges are of general concern, i. e. they are almost independent of the specific sensor application or the hydrogel type used and preclude the implementation of this sensor type in innovative real world applications in the first place (e. g. implantable sensors for continuous metabolic monitoring [10–12]).

Therefore, the aim of this work is the establishment of preferably general approaches for the improvement of piezoresistive hydrogel-based microsensors which can be similarly applied to different sensor types. In particular, a material-based approach concerning the deliberate modification of the hydrogel layer structure as well as the implementation of the force compensation method to hydrogel sensors are addressed (see Section 1.4).

1.2 Hydrogels and their applications

1.2.1 Definition, properties, and classification

A hydrogel can be understood as a two- or multicomponent system exhibiting a three-dimensional polymeric network structure. These structures can take up large amounts of a solvent and concurrently maintain their dimensional stability, i. e. they do not dissolve in the solvent but remain in a hydrated swollen state. In this sense, the word *solvent* is misleading and not, strictly speaking, correct but is well-established and, therefore, used in the following. One of the most important characteristics of a hydrogel is its permeability for water and its capacity of holding it. The three-dimensional integrity of hydrogels in the swollen state is caused by crosslinks – of either physical or chemical nature – between the polymer main chains. The elastic retracting force provided by the crosslinks balances the attempt of the polymer chains to dissolve. A force balance between these two opposing mechanisms leads to a steady state, the swelling equilibrium. The transition from a reference state to a new equilibrium state is a reversible process and referred to as volume phase transition (VPT). The amount of absorbed solvent can be tremendous and can even exceed the hundred-fold of the hydrogel's dry weight [13]. The concomitant presence of polymer chains (solid phase) and interstitial fluid (liquid phase) enables particular mechanical properties, i. e. visco-hyperelasticity or sometimes pure elasticity. The ability of solvent uptake depends, inter alia, on the equilibrium state of the chemical potentials between hydrogel and solvent, the polymer-solvent interaction, the presence of specific hydrophilic and hydrophobic side groups on the polymer backbone, and on the morphology of the network structure. Due to the diversity of existing polymeric hydrogel materials, they can be classified in a number of ways. A classification, considering important selected properties – focusing on hydrogels as sensor material – is given in Table 1.1.

As can be seen from Table 1.1, hydrogels can feature a variety of properties which can almost be combined arbitrarily. This enables fascinating materials which basically have the same structure but their properties can be tailored in a wide range with respect to specific applications.

The development of soft contact lenses based on acrylate gels by WICHTERLE in the 1960's can be considered as one of the first applications of hydrogels in the

Table 1.1: Classification of hydrogels.

Property	Feature	Type	Examples
Nature	natural		Alginate, Fibrin, Carrageenan
	synthetic		crosslinked poly(ethylene glycol) (PEG), poly(2-hydroxyethyl methacrylate) (PHEMA), derivatives of polyacrylamide (PAAm)
	hybrid		combinations of the above types
Response type	non-responsive		PEG, PHEMA
	stimuli-responsive		poly(*N*-isopropylacrylamide) (PNIPAAm), poly(acrylic acid) (PAA)
Ionic charge	uncharged	– non-ionic	(PNIPAAm), PAAm
	charged	– anionic	PAA
		– cationic	poly(*N,N*-dimethylaminoethyl methacrylate) (PDMAEMA)
		– ampholytic	combinations of anionic and cationic monomers
Crosslinking	physical		thermally cross-linked poly(vinyl alcohol) /poly(acrylic acid) (PVA/PAA)
	chemical		hydrogels cross-linked with *N,N'*-methylenebisacrylamide (BIS)
Morphology	non-porous		
	porous	– microporous	
		– mesoporous	
		– macroporous	
Composition	homopolymers		PVA, PAA
	interpenetrating polymers		PVA/PAA, PEG/chitosan
	copolymers	– triblock	poly(amidoamine)-poly(ethylene glycol)-poly(amidoamine)
		– multiblock	poly(ethylene glycol)-poly(amino-ester urethane)

(bio-)medical field [14]. Since that time, a stunning number of applications have been developed. Nowadays, hydrogels are used for emerging applications like:

Controlled drug delivery: A drug, incorporated inside the hydrogel matrix, is dissolved and released by diffusion when the hydrogel comes into contact with an aqueous medium. The hydrogel acts as permeable membrane or matrix for the drug [15].

Gene delivery: Hydrogels are used for non-viral incorporation of foreign DNA into a host cell and provide good transfection efficiency and biodegradability compared to other materials [16]. Silk-elastin like hydrogels have been reported to be very promising candidates for the treatment of solid tumors like human breast cancer [17, 18].

Wound dressings: Wound exudates as well as foreign bodies like bacteria are absorbed and retained by the hydrogel network. At the same time, the moisture content of the gel promotes fibroblast proliferation by reduction of the fluid loss from the surface of the wound as well as wound protection [19, 20].

Tissue engineering: Hydrogels are used to restore the function of traumatized or malfunctioned tissue or organs. Examples are three-dimensional corneal implantation, culturing of normal mature liver cells, or epithelial cell cultures of the breast [21, 22].

1.2.2 Hydrogels for sensor applications

The first reports on such hydrogels by KUHN [23, 24], BREITENBACH [25], and KATCHALSKY [26] date back to the end of the 1940's where polyelectrolytes were synthesized to a three-dimensional network . However, nearly no information was available on the internal thermodynamic processes and the gel-solvent interaction at that time. Thus, the postulate of a solvent-induced first-order volume phase transition by DUŠEK and PATTERSON in 1968 [27] was a remarkable statement, but could not be sufficiently confirmed in an experiment at that time. Another ten years later, TANAKA succeeded in the experimental evidence that a critical concentration of aceton in water leads to a collapse of a water-swollen acrylamide gel [28]. Consecutive studies could also prove a volume phase transition induced by other environmental parameters. This marked the beginning of intensive hydrogel research and led to an extensive development of sensitive hydrogels, responding to various kinds of stimuli. Nowadays, the variety of such gels is enormous, exhibiting response towards light (UV, IR), temperature, electric field, pH, carbon dioxide, ionic strength, concentration of organic co-solvents, mono- and divalent metal ions, glucose, antigens and many more [29]. Importantly, independent of the specific hydrogel, the volume phase transition is a reversible process enabling the use of these materials as sensitive sensor elements which are able to reversibly convert chemical energy into mechanical energy. The type of response can be systematically adapted by the "chemical design" of the hydrogel network, i. e. the selection of a specific matrix polymer and the incorporation of specific side groups to the polymer backbone. In this way, also multi-responsive gels can be synthesized. The choice of these reactive side groups determines the stimulus-response, i. e. whether the hydrogel responds to pH, glucose, antigens etc.

Generally, stimuli-sensitive hydrogels can be roughly classified into three categories, independent of the nature of the external stimulus: (I) stimulus-swelling, (II) stimulus-

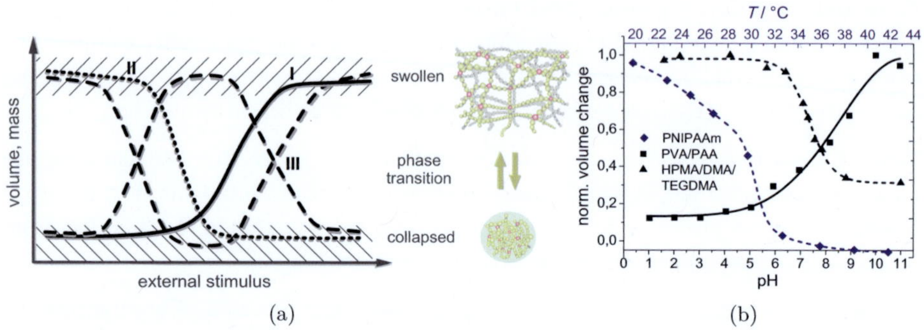

Figure 1.1: Volume phase transition of smart hydrogels: (a) stimulus-swelling (I), stimulus-shrinking (II), and stimulus-swelling-shrinking (III) smart hydrogels; (b) normalized volume change, experimentally determined from pH-responsive PVA/PAA hydrogels, poly(hydroxypropyl methacrylate-*N*,*N*-dimethylaminoethyl methacrylate-tetra-ethyleneglycol dimethacrylate) (HPMA/DMA/TEGDMA) hydrogels and temperature-responsive PNIPAAm hydrogels.

shrinking, and (III) stimulus-swelling-shrinking hydrogels according to their volume phase transition behavior (Fig. 1.1).
The exact characteristic of the swelling curve (shape, absolute volume change, intercept, threshold, saturation state, etc.) depends on a large spectrum of polymer-chemical quantities like monomer type, monomer concentration, ionic contribution of the reactive side groups, cross-linking density, and many more. The sigmoid shape of the characteristic curve, as depicted in Fig. 1.1, is most pronounced for pH-sensitive and temperature-sensitive hydrogels. For other hydrogels, the exact phase transition characteristics may significantly differ from a sigmoid shape. However, they still show monotonic increase, decrease or bijective behavior with increasing stimulus and can thus be generally assigned to one of the three classes. The large volume change in between the saturated states is explicitly used for sensor applications. However, the use of hydrogels with bijective volume phase transition behavior (curve III in Fig. 1.1a), either requires the limitation of the input range, so that only one flank of the curve is used during the measurement process, or the availability of an additional information from which flank the measurement information is obtained.

Despite the sensing properties of smart hydrogels, the mere material itself cannot be denoted as a sensor. In the general sensor definition, an analytical useful output signal (optical or electrical), which can be further processed, is presumed [30]. Therefore, additionally, a transducer, able to convert the volume change into such a signal, is required for the realization of hydrogel-based sensors.

1.3 State of the art of hydrogel-based sensors

1.3.1 Fundamental principle of hydrogel-based sensors

Hydrogels used for sensor applications are stimuli-responsive (cp. Table 1.1), i. e. small changes of an environmental trigger lead to large changes of the physico-chemical hydrogel properties. These gels are also referred to as stimuli-sensitive, smart, or intelligent hydrogels. Essentially, a hydrogel-based sensor is composed of two main components (Fig. 1.2): (I) a receptor part which is the stimuli-responsive hydrogel, acting as physico-/chemo-/bio-mechanical transducer, respectively, and (II) a transducer part which generates an optical or electrical output signal.

Hydrogel sensors based on free-swelling hydrogels take advantage of the altered physical properties of the gel network due to the volume change (ΔV). In contrast, another sensor type relies on mechanically confined hydrogel structures and analyzes the pressure change (Δp) exerted by the hydrogel. This volume or pressure change, induced by changes of the physical, chemical, or biological input quantity, is transformed into an appropriate analytical signal by the transducer. Direct hydrogel sensors, where the measurand directly affects the output signal, are most common. However, indirect methods, where the measurand affects an auxiliary quantity, e. g. pH, which in turn changes the output signal, as shown by HERBER *et al.* [31], do exist as well. This means that in such a case, the transducer part would comprise two or more transducers in series. In any case, it is assumed that the transducer itself shows no selectivity against the input quantity.

The unique possibility of a chemical design of the receptor part makes hydrogel-based sensors an astonishing sensor class which can selectively respond to a large variety of input signals by utilization of one and the same basic sensor principle. This can be rarely found for other sensor materials and is a remarkable advantage from a technological and economic point of view. For this reason, starting from the beginning of the 1990's, an increasing research interest regarding hydrogel-based

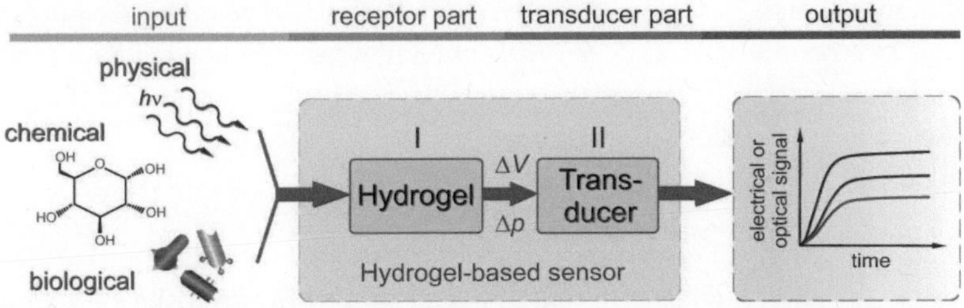

Figure 1.2: Fundamental principle of a hydrogel-based sensor comprising two main components: (I) a stimuli-responsive hydrogel as receptor part and (II) an arbitrary transducer.

sensors can be registered until today. One of the first works in the field of hydrogel-based microsensors was published by SHEPPARD in 1991 [32]. Since that time, various sensor types using different transducers have been developed for most diverse potential applications.

1.3.2 Conductometric sensors

Conductometric sensors account for a change in resistance when the input quantity is varied. In the work of SHEPPARD *et al.* [33,34], a planar platinum interdigitated electrode array (IDEA) structure covered with a 8 µm pH-sensitive poly(HEMA-*co*-DMAEMA) hydrogel layer is used. The volume change of the hydrogel induced by changes in the pH value leads to a changing conductivity of the hydrogel layer. The conductivity change and, thus, the pH value is measured by a change in impedance via the IDEA. The frequency range was chosen between 100 Hz and 100 kHz so that the impedance shows primarily resistive behavior and the imaginary part can be neglected. Due to the employed hydrogel, pH values between pH 7 and pH 9 can be measured. An increasing pH leads to a resistance increase due to the decreasing conductivity of the gel layer. An investigation of the adhesion behavior of photopatterned hydrogel structure on the substrate surface, using γ-methacryloxypropyltrimethoxysilane (γ-MPS) as adhesion promoter, revealed poor adhesion at high pH values (pH 9) and better adhesion at lower pH values (pH 6). However, it was found that drying of the hydrogel layer leads to a reformation of siloxane bonds between the gel and the substrate and, thus, can lead to improved adhesion. The influence of temperature (25 °C to 40 °C) on the adhesion properties was also studied, but found not to be of significant matter [35].

1.3.3 Optical sensors

Sensors based on optical transducers take advantage of different sensing mechanisms. They can directly measure the altered optical properties of the hydrogel (turbidity, refractive index) due to the volume phase transition. In a different approach, the movement of a certain filler material entrapped in a gel volume, a surface coating, or a periodic grating leads to changes of optical properties like a wavelength shift or the formation of a certain diffraction pattern.

a. Optical absorbance-based sensors

Some hydrogels show a change in optical transparency between the swollen and the shrunken state. The homogeneous network structure in the swollen state enables good transparency for visible light. The shrunken state is characterized by an

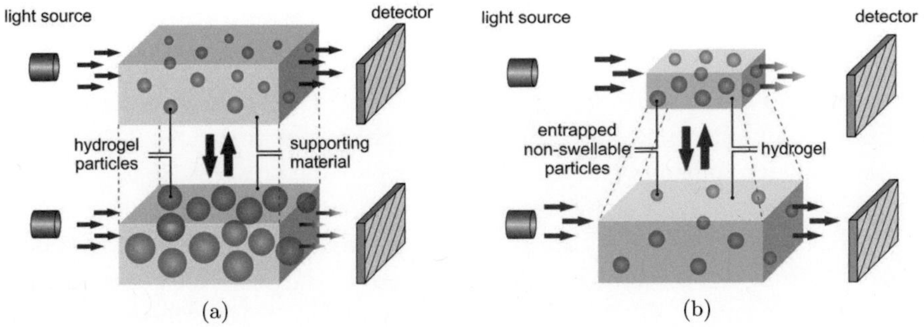

Figure 1.3: Hydrogel sensors based on optical absorbance: (a) swellable polymer microspheres entrapped in a non-swellable supporting material, and (b) non-swellable particles inclosed in a smart hydrogel matrix.

inhomogeneous distribution of polymer chains leading to opaqueness accompanied by a low optical transmission coefficient of the sensing element. Furthermore, the volume phase transition leads to a change in refractive index of the hydrogel. Hydrogels have a higher refractive index compared to water due to the presence of polymer chains. During the swelling process, the hydrogel's refractive index is decreased because of the increasing water content. The principle of hydrogel sensors based on optical absorbance is shown in Fig. 1.3.

The group of SEITZ reported on the incorporation of lightly cross-linked poly(vinylbenzyl chloride) (PVBC)-based microspheres into an acrylate-based and a PVA hydrogel membrane, respectively [36, 37]. The change in optical absorbance due to the volume change of the microspheres, initiated by the application of different pH values (pH 3 ... pH 9) as well as the concentration of divalent metal ions (Ca^{2+}, Cu^{2+}), has been measured with a conventional spectrometer. Microspheres made of different polymers inclosed in a PVA membrane have been also reported as optical sensing layers for Pb^{2+} and Hg^{2+} ions [38, 39].

A different approach, based on the measurement of the absorbance strength of localized surface plasmon resonance (LSPR), has been presented by ENDO *et al.* [40]. The proposed sensor measures the altered optical absorbance of LSPR due to changes of the interparticle distance of silver nanoparticles inclosed in the hydrogel due to hydrogel swelling/shrinking in the presence of glucose in a wide concentration range (10^{-13} M ... 10^{-3} M).

b. Diffractometric sensors

This sensor type is characterized by a shift in the diffraction wavelength or the change of a diffraction pattern during hydrogel swelling due to the movement of a periodic structure coupled to the hydrogel network.

An approach reported by the group of ZHANG [41] is based on a fiber BRAGG grating mechanically coupled to a responsive hydrogel layer (Fig. 1.4a). The injection of a spectrally broadband light signal into the optical fiber results in a reflection of light by the grating and in a wavelength shift when the pitch of the grating is varied. Variation of the grating period is accomplished by induced stress due to the volume phase transition of a PAAm hydrogel in response to different NaCl concentrations ($0.1\,mol/l \times 0.7\,mol/l$). The hydrogel pushes or pulls the clamps fixed on the fiber BRAGG grating which results in an increase or decrease in the BRAGG wavelength, respectively.

A crystalline colloidal array (CCA) structure that exhibits changes in its optical properties (colour change) in response to varying concentrations of metal ions (Pb^{2+}, Ba^{2+}, K^{+}) and glucose was first demonstrated by ASHER in 1997 [42]. In this approach, a three-dimensional crystalline structure is established by periodic assembly of highly charged polystyrene spheres in a stimuli-sensitive hydrogel matrix (Fig. 1.4b). Because the characteristic spacing between the spheres is in the order of hundreds of nanometers, a strong interaction with visible and infrared light is accomplished. Swelling or shrinking of the hydrogel matrix induced by changes of the analyte concentration change the intersphere distance resulting in an increase or decrease in the diffraction wavelength of the incident light and, thus, in a visible intense colour change. A change of 0.5 % of the hydrogel volume leads to a shift of the diffraction wavelength of about 1 nm [42]. For this approach, sensors with response to mono- and divalent metal ions, glucose, pH, organic solvents, different carbohydrates as well as organophosphorous compounds can be found in current literature [43–48]. Very recently, the group of JANA has shown a sensor for the detection of Hg^{2+} ions in aqueous media which employs an urease-immobilized CCA [49]. This sensor concept relies on a two-step-coupled spontaneous process of urea hydrolysis by urease and the inhibition of this hydrolysis by mercury ions leading to a very selective sensor device.

Similar ideas based on crystalline arrays have been implemented by other groups as well. In contrast to [42], BRAUN *et al.* used a polymeric crystalline colloidal array as a template, where the periodically assembled colloidal structure is dissolved after the surrounding hydrogel matrix is formed [50]. In this way, a light-diffracting hollow structure, also referred to as inverse opal, is established (Fig. 1.4b). The use of a poly(HEMA-*co*-AA) hydrogel for pH detection causes a red shift of the diffraction peak when the pH is increased. In [51] BRAUN *et al.* have shown that a rigorous optimization of different relevant technological parameters, like monomer ratio, polymerization time, optical filtering of UV light, and colloid material, leads to fast responding inverse opal sensors with a response time of about 12 s between pH

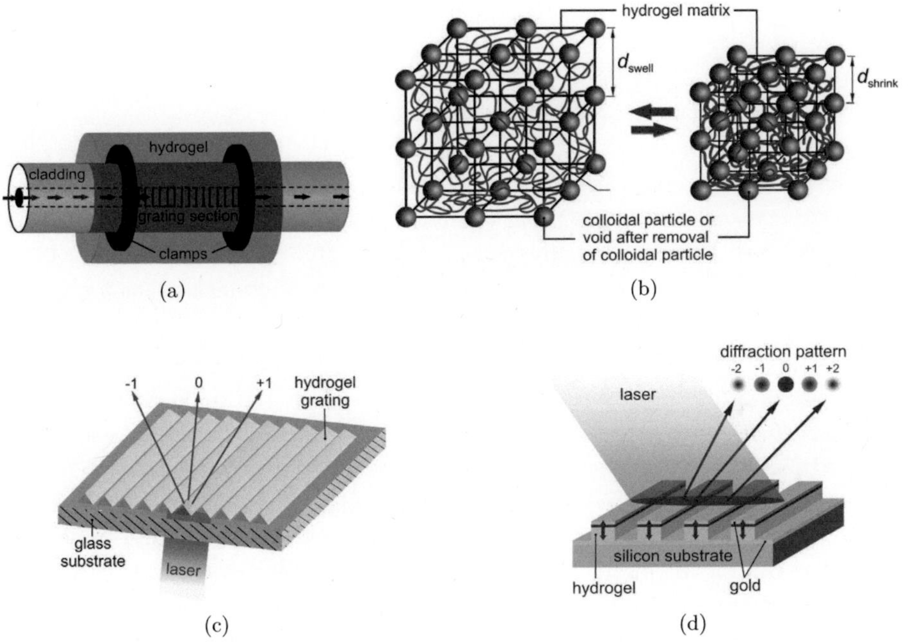

Figure 1.4: Optical diffractometric sensors based on hydrogel swelling: (a) fiber BRAGG grating coupled to a smart hydrogel layer, (b) crystalline colloidal array structure with regularly spaced particles or voids, respectively, in a hydrogel matrix, (c) swellable hydrogel grating attached to a glass substrate by microcontact printing, diffracts the incident laser light, (d) hydrogel-based diffraction grating with reflective gold coating.

5 and pH 6. This is an improvement of about two orders of magnitude compared to sensors reported earlier [50]. In [52] a new and more general strategy for the enhancement of the sensitivity and response time of an inverse opal sensor has been shown by the introduction of a planar defect layer. A three-dimensional macroporous array of poly(methacrylic acid)(PMAA) hydrogel, which contains a planar defect layer of differently sized macropores, was formed by a combination of photonic crystal-templating method and LANGMUIR-BLODGETT technique and was investigated by reflection spectroscopy. For the sensor containing a defect layer compared to a sensor without defect layer, a sensitivity enhancement of about 50 % and a response time reduction of circa 50 % was reported for a pH range between pH 7.5 and pH 8.5. The enhanced response time for swelling and deswelling was measured to be 9 s and 13 s, respectively, which is in the same range as the results shown by BRAUN in [51].

In the approach of ASHER *et al.* [42], a very high sensitivity is obtained because large domains of single crystals are formed which leads to an intense and narrow

BRAGG diffraction band. However, the hydrogel matrix has to be formed without disturbing the periodic structure of the charged spheres, i. e. charged hydrogels can not be used, which limits the variety of potential target analytes. Furthermore, this CCA exhibits a rather large diffusion barrier for the target analytes leading to long response times. In contrast, the inverse opal structure by BRAUN *et al.* [50] offers more flexibility in terms of the functional groups and charges. The interconnected porous structure allows a rapid transport of the analyte solution and enables a faster sensor response. However, a reduced mechanical stability, as a consequence of the porous structure, has to be accepted and considered as crucial factor for specific applications.

The approach by the group of LOWE is based on a holographic diffraction grating attached to glass or plastic substrates [53, 54]. Here, lines of spherical silver nanoparticles with a diameter of less than 20 nm are regularly arranged in a smart hydrogel matrix by a photosensitization process using a frequency-doubled Nd-YAG laser. After the formation, the silver particles lay in parallel planes, also referred to as fringes. During the illumination of the holographic diffraction grating by white light, it acts as sensitive wavelength filter. The reflected narrow band of wavelengths depends on the grating spacing which is altered by the volume phase transition of the hydrogel matrix. Constructive interference of the reflections from each fringe plane results in a characteristic spectral peak which can be used for sensor calibration. This holographic principle was applied to sensors for the measurement of mono- and divalent metal ions, pH, ionic strength, glucose, penicillin, urea, bacterial spores [55–61] as well as enzyme inhibition assays [62]. The concept of LOWE's group has been adopted by PRITCHARD *et al.* for the *ex vivo* real time monitoring of pH and glucose in human blood plasma [63, 64]. From a practical point of view, the use of a silver halide photographic emulsion in combination with the laser to fabricate the holographic structure can be considered the most challenging step of this approach.

A different sensor concept, which employs a structured hydrogel layer itself as diffraction grating without the need of an additional periodic structure, like the approaches presented above, was described in [65]. Here, a periodic array of hydrogel microstructures has been fabricated by UV-assisted micro-imprinting using a polydimethylsiloxane (PDMS) master mold. Two types of gratings, i. e. a hydrogel grating attached to a glass substrate (Fig. 1.4c) and a sheet-like free-standing structure (not shown), respectively, have been presented. The illumination of the hydrogel grating results in a diffraction of the incident light and in a defined diffraction pattern. Swelling or shrinking of the hydrogel structure in presence of a certain pH value changes the periodicity of the grating and, thus, the properties of the diffraction pattern, i. e. intensity, spot size, and position. These property changes are usually already detectable with the bare eye.

A similar approach was presented by the group of WANG, which also used a soft-lithographic stamping technique for the fabrication of hydrogel gratings functionalized with glucose oxidase (GOx) and phenylboronic acid groups, respectively, for glucose monitoring [66,67]. The grating period was reported to be in the order of $(1 \ldots 4)\,\mu m$. In [68] a 2D grating for pH sensing, based on an acrylic acid hydrogel, was presented.

Very recently, a reflective diffraction grating made of smart hydrogels, where a laser beam is reflected from the hydrogel grating surface coverd with a gold layer, was reported by Chang *et al.* [69]. Here, hydrogel swelling/shrinking in response to pH leads to a height change of the grating and, therefore, to a change of the diffraction intensity of the reflected light (Fig. 1.4d). In this approach, the incident light only interacts with the gold surface but not with the hydrogel material itself. Since the width and the height of the hydrogel stripes are in the sub-10 µm range, even small height changes in the sub-µm range can be detected. This makes the device very sensitive compared to other approaches.

c. Interferometric sensors

Sensors where a hemispherical smart hydrogel is attached to the tip of an optical fiber, containing a FABRY-PEROT cavity, have been presented by TIERNEY *et al.* [70] (Fig. 1.5a). The injected light is reflected at the fiber-gel interface and the gel-solution interface generating an interference pattern inside the fiber. The phase of the interference pattern depends on the optical path length inside the hydrogel structure and, thus, on the swelling state of the gel. Therefore, a volume change of the gel due to changes in the analyte concentration alters the distance between the fiber-gel and the gel-solution interface and effects the phase of the interference pattern. Here, acrylamide-based hydrogels with a radius in the order of 50 µm have been used. With this technique even a displacement of 2 nm could be detected, revealing that this method can achieve high resolution and good sensitivity. The group of TIERNEY has reported interference-based sensors for the measurement of ionic strength, pH, glucose, and different carbohydrates as well as concentration and sequence of short nucleotides [70–72].

d. Reflective sensors

This kind of hydrogel sensors are similar to absorbance-based sensors described in Section 1.3.3a. However, not the transmitted intensity through a membrane is measured, but the intensity of light reflected by a porous swellable polymer attached to an optical fiber as shown in [73]. A pH-sensitive porous aminated polystyrene

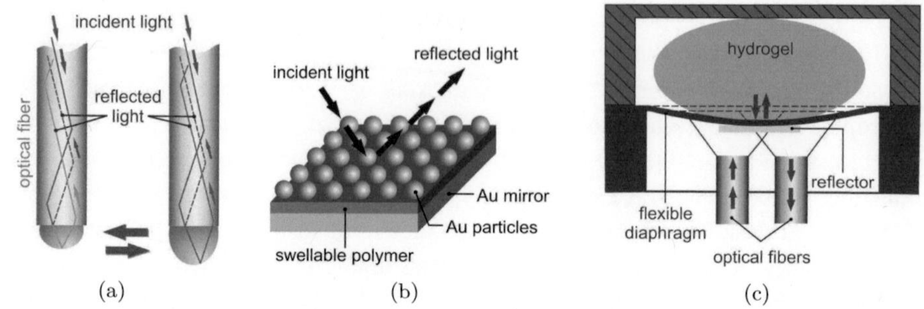

Figure 1.5: Interferometric and reflective sensors based on hydrogels: (a) hemispherical hydrogel attached to the tip of an optical fiber, containing a FABRY-PEROT cavity, (b) optical thin-film system where the smart hydrogel layer containing gold metal islands on its surface is attached to a reflective surface (mirror), (c) smart hydrogel structure coupled to a flexible diaphragm containing a reflector.

polymer increases its clarity due to swelling accompanied by a decrease in the reflected intensity. Because of the porous structure, diffuse reflection occurs at the polymer-pore interfaces due to different refractive indices of the polymer and the solution inside the pores. In the pH range between pH 6.5 and pH 8, an intensity decrease of about factor two has been reported [73]. The optical system comprises a LED as light source, a photodiode detector, and a fiber-optic coupler serving as beam splitter. Furthermore, transducers based on a bundle of 20 bifurcated optical fibers and an optical reflective device have been shown [74, 75].

Another approach is based on an optical thin-film system. A responsive hydrogel layer is attached to a mirror and contains gold metal islands on its top surface (Fig. 1.5b). This stacked assembly acts as optical reflection interference filter whose reflection output signal depends on the hydrogel layer thickness. This layer thickness is modulated by pH-induced swelling and is analytically evaluated by monitoring the slope of the characteristic spectral reflection minimum [76, 77].

In an optical reflection principle relying on mechanical work of the hydrogel, the polymer is mechanically coupled to a reflective diaphragm (Fig. 1.5c). A beam of light is sent to the reflective plane using an optical fiber. The volume phase transition of the hydrogel leads to the deflection of the reflective diaphragm and to a change in the intensity of the light reflected back into a second optical fiber [78, 79].

e. Energy transfer-based sensors

In general, two energy transfer-based techniques, i. e. nonradiative energy transfer (NRET) [80] and fluorescent energy transfer (FRET) [81], are known for the detection of volume changes. The attachment of donor-acceptor pairs to smart hydrogels enables the detection of the volume phase transition by measuring the relative intensity of each dye which depends on the donor-acceptor distance. In [81,82] FRET fluorophore pairs have been incorporated in pH-sensitive chitosan hydrogel microspheres and loaded with GOx to respond to glucose.

1.3.4 Resonance frequency-based sensors

Resonance frequency-based hydrogel sensors use transducer types that change their resonance frequency depending on a certain change in mass load, i. e. mass increase or decrease.

a. Quartz crystal microbalance

In quartz crystal microbalance-based sensors, an alternating electric field stimulates the vibration of a quartz crystal to oscillate in the resonance frequency. Loading of this quartz resonator with a hydrogel and initiation of a volume phase transition by a certain analyte results in a change of this load and, consequently, in a shift of the resonance frequency. Not only the mass change of the gel during swelling/shrinking will influence the resonance frequency shift but also the changing viscoelastic material properties. The latter affects the decay rate of the oscillations. Hydrogel-coated quartz crystal microbalance resonators have been reported for the detection of different nucleotides, pH, antibodies and, very recently, for Cu^{2+} ions in aqueous solutions [83–86]. Because frequency can be measured very precisely, this sensor type has high sensitivity. Due to the thin hydrogel layer, usually with a thickness in the sub-µm range, response times are relatively small. However, the uniformity of the gel layer, adhesion, and miniaturized sensors are major concerns regarding their application.

b. Magnetoelastic

Magnetoelastic sensors have been mainly studied by GRIMES *et al.* and use a magnetoelastic ribbon coated with a stimuli-sensitive hydrogel layer [87]. This structure is excited by a time-varying magnetic field using a drive coil which induces a mechanical oscillation of the sensing element and, thus, a magnetic flux. The transient response is captured remotely by a sensing coil and the resonance frequency is determined either by frequency counting or fast Fourier transform (FFT) of the response signal. The swelling/shrinking of the hydrogel layer changes the resonance frequency. Similarly to the quartz crystal microbalance, the resonance frequency is also altered by other factors like elasticity of the gel layer, temperature as well as viscosity and density of the analyte solution. Beside the use of these sensors for pH measurement, they have been used for the detection of organophosphorus pesticides by the same group [88]. These sensors benefit from the wireless interrogation but suffer from poor adhesion of the polymer layer on the metallic, magnetoelastic ribbon.

1.3.5 Mechanical sensors

Mechanical hydrogel sensors directly use the mechanical work, evolved from the volume phase transition, to induce strain or deformation to a mechanical transduction element, like a flexure beam or a plate, which can be detected.

a. Microcantilever-based sensors

This approach relies on the ability of microcantilevers to bend if they are subjected to a stress state. If the top or the bottom side of the cantilever beam is coated with a hydrogel layer, hydrogel swelling/shrinking leads to a lateral surface stress state and, consequently, provokes bending of the microcantilever (Fig. 1.6a). The bending state can be detected for example by a four quadrant photodetector which records the position of a laser beam reflected from the cantilever tip [89]. In this case, the tip of the cantilever cannot be coated with hydrogel to enable proper reflection of the laser beam. The latter method has been mainly used by the groups of JI and PEPPAS for the detection of CrO_4^{2-}, Pb^{2+}, Be^{2+}, glucose, and pH [90–96]. A second common method – beside optical readout – is the integration of piezoresistors inside the cantilever material near to the fixed support and their connection in a WHEATSTONE bridge configuration. Here, the resulting voltage change due to resistivity change of the piezoresistors is correlated to the swelling state of the hydrogel [89]. This method has been used by the group of URBAN for measurement of pH [97].

b. Capacitive flexure plate transducers

Capacitive transducers together with piezoresistive pressure transducers, described in the next section, can be classified as flexure plate transducers. The volume phase transition of the hydrogel leads to a contact pressure on the bending plate which is fixed on all edges. For capacitive transducers, the bending plate acts as flexible electrode. The distance between this electrode and an opposing planar fixed electrode is altered depending of the hydrogel's degree of swelling and results in a capacitance change (Fig. 1.6b). The integration of the variable capacitor to a resonant LC circuit with a fixed inductance leads to a shift of the resonance frequency which can be measured precisely and even wirelessly by inductive coupling with an external coil [98]. Nevertheless, regarding miniaturization, the required space for a coil, even if manufactured with micromaching techniques, is rather large and determines the

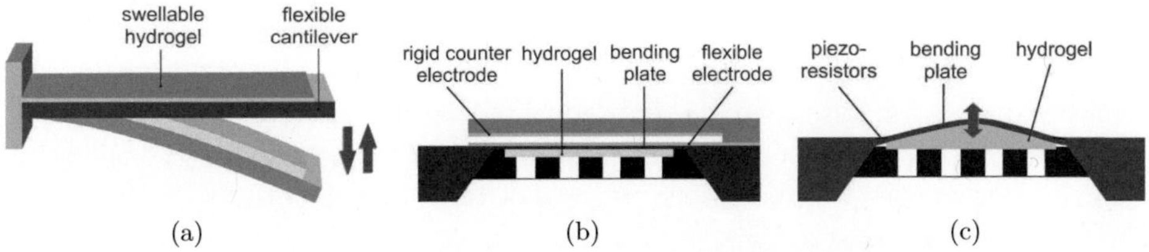

Figure 1.6: Mechanical hydrogel-based sensors: (a) microcantilever-based hydrogel sensor, (b) capacitive flexure plate transducer, and (c) piezoresistive flexure plate transducer.

dimensions of the final device. Capacitive implantable micromachined hydrogel sensors for metabolic monitoring of glucose and pH have been primarily advanced by the SIEGEL group [99–101].

c. Piezoresistive flexure plate transducers

This class of hydrogel sensors directly converts the mechanical work performed by the hydrogel into an electrical output signal using the piezoresistive effect. In particular, pressure sensors with integrated piezoresistors in WHEATSTONE bridge configuration at the edge of a thin flexible bending plate are used as transducer elements. The hydrogel itself is encapsulated in the cavity on the back side of the bending plate (Fig. 1.6c). In this way, free swelling is deliberately constrained and the gel structure exerts a pressure which can be measured. Existing approaches from different groups mainly differ in the specific design of the pressure sensor chip, the supply of the measuring solution, and the type, attachment, and structure of the sensitive hydrogel layer.

The group of GERLACH basically reported on two different sensor types. The first design is characterized by a comparatively thick hydrogel layer (50 µm ... 90 µm) which enables higher sensitivity but also shows longer response times. In this design, the hydrogel is not mechanically constrained by the side walls of the pressure sensor chip but only by the bending plate and a rigid support at the bottom of the gel layer. The second type exploits the bimorph effect by deposition of a thin hydrogel film (5 µm ... 50 µm) onto the backside of the bending plate. This design is advantageous regarding small response times due to the thin hydrogel layer. However, the work performed by the bimorphic approach is much smaller compared to volumetric hydrogel swelling resulting in a smaller sensitivity [6, 102]. In both designs, fluidic channels are used for supplying the measuring solution by a pump. The same group reported on the simulation of these sensors by the finite element method (FEM) and the description of hydrogel swelling inside the sensor by the MOONEY-RIVLIN equation [7]. In [103] chemical sensors based on multi-responsive hydrogels, i. e.

ionic co- and terpolymers of PNIPAAm, have been presented for the measurement of pH, divalent transition-metal ions as well as different salt concentrations and the conditions for long-term stability and high signal reproducibility were stated. Further work has been done on modeling of non-linear effects in hydrogel swelling and comparison to experimental results to get a deeper understanding of the underlying phenomena and, thus, to be able to improve the sensor performance [104].

HAN *et al.* followed a slightly different approach where the hydrogel is completely enclosed inside the cavity of the pressure sensor. The sensor is closed from the back side by a rigid porous membrane for supplying the measuring solution. In this case, the hydrogel can only swell in vertical direction towards the flexure plate and is mechanically constrained from all other sides. Because of the relatively small volume change, they call the device a "constant volume hydrogel osmometer" even if, strictly speaking, the volume is not kept constant [105]. This approach was recently revived by the group of SOLZBACHER which intends to build a hydrogel-based pressure sensor array for continuous *in vitro* metabolic monitoring of diabetes patients [11]. For this purpose, they fabricated a micromachined 2 x 2 pressure sensor array where the flexure plate is perforated by circular holes in a butterfly-like structure. In this design, a porous backplate is no longer needed because the analyte solution can be directly applied to the perforated front side. However, a decrease in sensitivity of about 16 % to 34 % for different hole diameters (10 µm ... 40 µm) has been found compared to a solid flexure plate [106]. These sensor arrays have been also tested in *in vitro* experiments for measurement of ionic strength of PBS model solutions and compared to sensor arrays with solid diaphragm and solution supply from the sensor back side [12]. In [10] different two-layer combinations of wiremeshs and porous membranes have been used as materials for the backplate and their influence on sensor sensitivity and response time has been systematically studied. Furthermore, the impact of the initial pressure on the magnitude of the sensor response was studied. An interesting result revealed that a higher sensitivity could be achieved by application of an increased loading pressure.

HERBER *et al.* reported on a two-step method where the pH of an electrolyte solution is changed by CO_2 partial pressure (pCO_2). This change in pH of the electrolyte is measured by a pH-sensitive hydrogel inside a pressure sensor. The resulting output voltage was then correlated to pCO_2, relevant for luminal CO_2 measurement for early recognition of gastrointestinal ischemia [31].

1.3.6 Summary

For comparison, selected parameters as well as the employed hydrogels of different hydrogel-based pH sensors, as presented in the previous sections, are summarized in

Table 1.2.

Table 1.2: Selected parameters of different hydrogel-based pH sensors (* values have been estimated from experimental curves).

Sensor type	Hydrogel	Layer thickness in µm	pH range	Sensitivity per pH unit	Response time in s	Ref.
Conductometric	poly(HEMA-*co*-DMAEMA		7...8	100 Ω	350	[34]
Optical absorbance	PVBC with diethyl malonate	few	4.8...7.7	0.05 absorbance units	300	[37]
Diffractometric	PAAm-PCCA	125	4.3...8.5	74.2 nm	600	[44]
	poly(HEMA-*co*-AA)	25	5...6	80 nm	12	[51]
	PMAA	2.8	7.5...8.5	60 nm	9	[52]
	poly(HEMA-*co*-MAA)	10	4...6	125 nm	250	[56]
	poly(AAm-*co*-MAA)	2	4.6...5	4 µm	300*	[69]
Interferometric	poly(AAm-*co*-DMAPAA)	60	8...9	45 µm*	95	[70]
Quartz crystal microbalance	PVA/PAA	0.39	2.55...3.45	13.2 kHz	0.5	[84]
Magnetoelastic	poly(AA-*co*-IOA)	1.36	4.4...8.5	0.506 kHz	120	[87]
Microcantilever	poly(HEMA-*co*-DMAEMA)	2	N. A.	N. A.	600	[97]
	PMAA	3 µm	5...6.5	17.5 µm	N. A.	[96]
Capacitive	poly(AAm-*co*-MAA)	200	3...6.5	1.16 MHz	2700	[100]
Piezoresistive	PVA/PAA	40	5.5...11	14.5 mV*	720	[7]
	poly(HEMA-*co*-DMAEMA)	15	6...7.5	60 kPa	600	[107]
	PVA/PAA	6	3...4	2.5 mV	90	[108]
	P2VP-block-poly(NIPAAm-*co*-DMIAAm)	4	2...5	12.5 mV	70	[103]

PCAA polymerized crystalline colloidal array, DMAPAA *N*-(3-dimethylaminopropyl)acrylamide, P2VP poly(2-vinylpyridine), DMIAAm (2-dimethyl maleinimido)acrylamide

1.4 Scope of this work

This work goes along with intensive research work performed on hydrogel-based sensors at the Solid-State Electronics Laboratory (IFE), TU Dresden, in the last decade [8, 109] and builds on the profound knowledge developed there.

The particular scope of this work is the improvement of piezoresistive hydrogel-based sensors using flexure plates with particular respect to sensor response time, suppression of dynamic drift effects as well as mechanical hydrogel stability which can be considered as serious drawbacks regarding the application of these sensors. The designated aim is the establishment of general improvement approaches which cannot only be applied to the particular sensor set up studied here, but also to other piezoresistive sensors using different hydrogels.

Based on the fundamental principle of hydrogel-based sensors which are composed of a receptor part (hydrogel layer) and a transducer part (flexure plate with piezoresistors; see Fig. 1.2), in general, the sensor properties can be improved by:

(i) altering the transducer properties,
(ii) altering the hydrogel layer properties, and
(iii) the application of advanced measurement methods.

Since piezoresistive transducers are under development for more than four decades, they are highly developed and the sensor properties to be improved primarily depend on the properties of the hydrogel material. Therefore, the first approach is less promising but the other two are considered within this work. Therefore, the thesis is divided in two main parts:

1. *Hydrogels with improved material properties* (Section II): Here, acrylamide-based hydrogels used as sensing layers will be modified in order to attain improved sensor properties.

    - As a first aim, the long response times of pH-sensitive ionic copolymers of acrylamide, i. e. poly(acrylamide-*co*-acrylic acid) (p(AAm-*co*-AA)), are intended to be reduced by the incorporation of an interconnected pore structure with pores in the µm-range. This is supposed to lead to: (i) a drastic increase in the surface-to-volume ratio to increase the surface area between the solvent and the hydrogel and to improved solution transport into the gel by convection through the pores rather than by diffusion, (ii) a reduced critical hydrogel dimension (wall between pores) which determines the slow cooperative diffusion process (Chapter 3).
    - A second aim is the improvement of the mechanical stability of non-ionic poly(*N*-isopropylacrylamide) (PNIPAAm) hydrogels, sensitive to temper-

ature, organic co-solvents and ionic strength, in order to improve the long-term stability and the repeatability of the sensor signal. Here, the fabrication and integration of physical hydrogel networks, using nanoparticles as multifunctional crosslinks between polymer chains, is pursued (Chapter 4).

2. *Hydrogel-based sensors with force compensation as measurement method* (Section III): Here, for the very first time, the force compensation method is applied to hydrogel-based sensors to maintain a constant hydrogel volume during the whole measuring process and, thus, to suppress the effect of cooperative diffusion and dynamic drift effects.
   - At first, a theoretical system-based description of the force compensation method compared to the deflection method shall elucidate the advantages of the compensation method with respect to important figures of merit in sensor technology, like response time, drift suppression, and suppression of systematic static deviations (Chapter 5).
   - Secondly, a hydrogel-based sensor which can be operated with the force compensation method shall be designed, fabricated and studied and the measurement results shall be compared to conventional sensors operated in deflection method (Chapter 6).

Finally, in Chapter 7, the essential results are summarized and an outlook with respect to future work is given in Chapter 8.

2 Theoretical foundations

Within this chapter, the theoretical background for understanding the behavior of the two main sensor components, i. e. the sensitive hydrogel layer and the piezoresistive pressure transducer, as well as the underlying physical and chemical phenomena are presented. At first, the driving force and the process of reversible volume phase transition of stimuli-sensitive hydrogels, as briefly outlined in the previous chapter, is theoretically described by means of thermodynamic considerations. Furthermore, the phenomena influencing the kinetics of the swelling process and, thus, directly influencing the sensor response time as well as common mechanical models for the macroscopic description of the viscoelastic gel properties are discussed. The second part of this chapter deals with a brief introduction to piezoresistive pressure transducers. Subsequently, the influence of the geometric properties of the bending plate on the transducer behavior is presented and discussed.

2.1 Hydrogel swelling

The theory described in the following assumes isobaric conditions and, therefore, holds for the scenario of free swelling. A more detailed description can be found in [110, 111].

2.1.1 Thermodynamics of swelling equilibrium

The theoretical description of swelling equilibrium is based on a thermodynamic model developed by FLORY [110]. Here, the description of thermodynamic processes is carried out by means of the chemical potential μ_i which can be written as

$$\mu_i = \left(\frac{\partial G}{\partial n_i}\right)_{T,p,n_{j\neq i}}, \tag{2.1}$$

where G denotes GIBBS free energy and n_j the number of moles of all system components except n_i.

If a hydrogel is subjected to a certain solvent, a difference in chemical potential between the solvent inside the hydrogel and the solvent in the surrounding solution exists and is accompanied by an osmotic swelling pressure

$$\Pi_{\mathrm{swell}} = -\frac{1}{V_{\mathrm{s,m}}} \cdot \left(\frac{\partial \Delta G_{\mathrm{tot}}}{\partial n_{\mathrm{s}}}\right)_{T,p} = -\frac{(\mu_{\mathrm{s}} - \mu_{\mathrm{s}}^{0})}{V_{\mathrm{s,m}}} = -\frac{\Delta\mu}{V_{\mathrm{s,m}}}. \tag{2.2}$$

Here, $V_{\mathrm{s,m}}$ is the molar volume of the solvent, ΔG_{tot} the total change of GIBBS free energy of the gel network, n_{s} the number of moles of the solvent, μ_{s} the chemical potential of the solvent inside the hydrogel, and μ_{s}^{0} the reference chemical potential of the solvent in the surrounding solution. From Eq. (2.2) it becomes clear that a difference in chemical potential is the driving force for a VPT of the hydrogel (solvent uptake or release of the gel network) and that an osmotic swelling pressure (or swelling force) is generated as long as the network is not yet in its equilibrium state. In swelling equilibrium, the difference in chemical potential as well as the osmotic swelling pressure equal zero ($\Delta\mu = 0, \Pi_{\mathrm{swell}} = 0$).

The FLORY-REHNER theory combines the FLORY-HUGGINGS theory for polymer solutions and the rubber elasticity theory (RET) and acts on the assumption of a linear superposition of both contributions. In general, the network deformation can be regarded as the result of the superposition of two physico-chemical processes: (i) the mixing of polymer and solvent and (ii) the elastic deformation of the network. Thus, the total change in GIBBS free energy of the network can be expressed as the sum of the changes in GIBBS free energy of the single processes as

$$\Delta G_{\mathrm{tot}} = \Delta G_{\mathrm{m}} + \Delta G_{\mathrm{el}} = \sum_{\nu=1}^{2} \Delta H_{\nu} - T\Delta S_{\nu}, \tag{2.3}$$

where ΔG_{m} and ΔG_{el} are the changes in GIBBS free energy due to mixing and elastic deformation, respectively, T is the temperature, and ΔH_{ν}, ΔS_{ν} are the enthalpy change and the entropy change of the particular contribution, respectively.

Analogously, the total osmotic pressure can be written as

$$\Pi_{\mathrm{swell}} = \Pi_{\mathrm{m}} + \Pi_{\mathrm{el}}. \tag{2.4}$$

a. Swelling of neutral hydrogels

If a polymer network is brought into contact with a solvent, the mixing contribution results from penetration and distribution of solvent molecules in the network, which increases the entropy of the network. At the same time, heat transfer between the solvent and the gel is assumed and can be described by a change of enthalpy. Therefore, the change in GIBBS free energy due to this mixing process - according to the FLORY-HUGGINS theory - can be expressed as

$$\Delta G_{\mathrm{m}} = \Delta H_{\mathrm{m}} - T\Delta S_{\mathrm{m}} = RT(n_{\mathrm{s}} \cdot \ln\varphi_{\mathrm{s}} + n_{\mathrm{p}} \cdot \ln\varphi_{\mathrm{p}} + \chi n_{\mathrm{s}}\varphi_{\mathrm{p}}), \tag{2.5}$$

where R is the universal gas constant, n_s, n_p, and φ_s, φ_p are the number of moles and the volume fraction of solvent and polymer, respectively, and χ is the FLORY-HUGGINS interaction parameter, which describes the affinity of the solvent to mix with the polymer chains. The smaller χ is, the better is the quality of the surrounding solvent to mix with the polymer chains.

Considering

$$\varphi_s = \frac{n_s}{n_s + sn_p}, \quad \varphi_p = \frac{sn_p}{n_s + sn_p} \tag{2.6}$$

with s as the number of segments of the polymer chains and

$$\varphi_s = 1 - \varphi_p, \tag{2.7}$$

together with Eq. (2.2) yields

$$\Pi_m = -\frac{\Delta\mu_m}{V_{s,m}} = -\frac{1}{V_{s,m}}\left(\frac{\partial(\Delta G_m)}{\partial n_s}\right) = \frac{RT}{V_{s,m}}\left[\ln(1-\varphi_p) + \varphi_p + \chi\varphi_p^2\right]. \tag{2.8}$$

This equation is valid under the assumption that the gel network is one macromolecule of infinite molecular weight ($s \to \infty$). The volume fraction of polymer φ_p is the ratio of polymer volume V_p to the total volume of polymer and solvent ($V_p + V_s$) and is, thus, inversely proportional to the volume degree of swelling Q_v of the hydrogel as

$$\varphi_p = \frac{V_p}{V_p + V_s} = Q_v^{-1}. \tag{2.9}$$

The solvent uptake of the gel network due to the contact between hydrogel and solvent and the difference in chemical potential between both phases results in an isotropic stretch of the network chains. Due to the existence of crosslinking points, the chains cannot dissolve in the solvent and stretching consequently leads to an elastic retracting force which counteracts swelling. This phenomenon is well known from rubber-like materials under tensile deformation. According to FLORY, the heat change during this process is negligible ($\Delta H_{el} = 0$) and the only phenomenon which contributes to a change in GIBBS free energy is the change of entropy due to chain stretching, i. e. the isotropic transition from volume V_p to the new volume $V = V_p + V_s$, which is Q_v-times larger than V_p . Thus, GIBBS free energy due to network deformation can be written as

$$\Delta G_{\mathrm{el}} = -T\Delta S_{\mathrm{el}} = \frac{kTN_{\mathrm{c}}}{2}\left(3\lambda^2 - 3 - \ln\lambda^3\right), \tag{2.10}$$

where k denotes the BOLTZMANN constant, N_{c} the number of polymer chains, and λ the isotropic expansion factor. The above equation holds for the assumption of isotropic network deformation, i. e. $\lambda = \lambda_{\mathrm{x}} = \lambda_{\mathrm{y}} = \lambda_{\mathrm{z}}$, and a GAUSSian distribution of the end-to-end distance of the polymer segments between two junction points. λ can be expressed by the polymer volume fraction φ_{p} or the volume degree of swelling Q_{v} as

$$\lambda^3 = \lambda_{\mathrm{x}}\lambda_{\mathrm{y}}\lambda_{\mathrm{z}} = \frac{V}{V_{\mathrm{p}}} = \frac{V_{\mathrm{p}} + n_{\mathrm{s}}V_{\mathrm{s,m}}}{V_{\mathrm{p}}} = \frac{1}{\varphi_{\mathrm{p}}} = Q_{\mathrm{v}} \tag{2.11}$$

Derivation of Eq. (2.10) with respect to n_{s} and taking into account Eq. (2.11) leads to the osmotic pressure due to elastic network deformation:

$$\Pi_{\mathrm{el}} = -\frac{\Delta\mu_{\mathrm{el}}}{V_{\mathrm{s,m}}} = -\frac{1}{V_{\mathrm{s,m}}}\left(\frac{\partial\Delta G_{\mathrm{el}}}{\partial\lambda}\right)\left(\frac{\partial\lambda}{\partial n_{\mathrm{s}}}\right) = -\frac{kTN_{\mathrm{c}}}{V_{\mathrm{p}}}\left(\varphi_{\mathrm{p}}^{1/3} - \frac{\varphi_{\mathrm{p}}}{2}\right). \tag{2.12}$$

To obtain a more meaningful expression in terms of application, the number N_{c} of polymer chains can be expressed in terms of the cross-linking density ν_{c} as

$$N_{\mathrm{c}} = \nu_{\mathrm{c}}V_{\mathrm{p}}N_{\mathrm{A}}, \tag{2.13}$$

with AVOGADRO number N_{A}, leading to

$$\Pi_{\mathrm{el}} = -RT\nu_{\mathrm{c}}\left(\varphi_{\mathrm{p}}^{1/3} - \frac{\varphi_{\mathrm{p}}}{2}\right). \tag{2.14}$$

Eqs. (2.12) and (2.14) are valid for the assumption of perfect tetra-functional networks, which is a sufficient approximation for the considerations within this work. However, according to [112], the behavior of a real network is considered to reside between two limits, corresponding to the classical approaches of the JAMES-GUTH phantom network theory [113], i. e. free-fluctuation limit (localized phantom network) and the limit of full suppression of junction fluctuations (affine network). In this case, the change in GIBBS free energy also depends on the so-called cycle rank and the functionality of crosslinks, which in turn depends on their chemical nature.

The total swelling pressure of neutral hydrogels can be written as

$$\Pi_{\mathrm{swell}} = RT\left\{\frac{1}{V_{\mathrm{s,m}}}\left[\ln(1-\varphi_{\mathrm{p}}) + \varphi_{\mathrm{p}} + \chi\varphi_{\mathrm{p}}^2\right] + \nu_{\mathrm{c}}\left(\varphi_{\mathrm{p}}^{1/3} - \frac{\varphi_{\mathrm{p}}}{2}\right)\right\}. \tag{2.15}$$

b. Swelling of ionic hydrogels

For networks containing ionic groups (e. g. pH-sensitive hydrogels), the contribution of such ions as well as their counterions have to be considered in addition to the contributions to hydrogel swelling considered so far. Thus, Eqs. (2.3) and (2.4) must be extended by adding ΔG_{ion} and Π_{ion}, respectively. The presence of ionic groups results in an enhanced swelling force due to the interaction between the polymer network and the respective ionic species. Based on the DONNAN theory [114, 115], the charge difference between the gel and the solvent phase and the concomitant exchange of ions between the swollen gel and the solution is considered (Fig. 2.1).

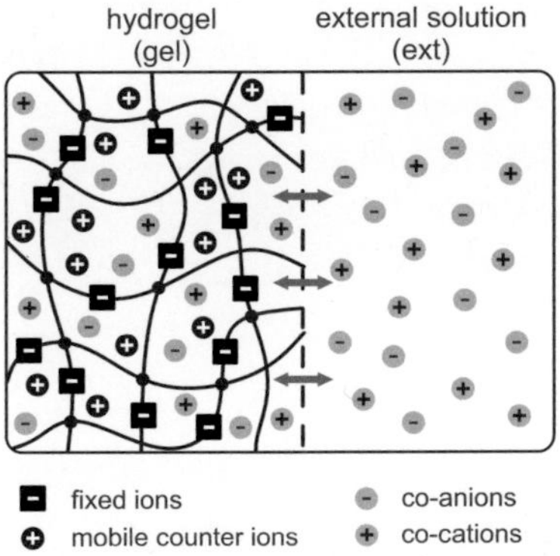

Figure 2.1: Simplified schematic illustration of an anionic hydrogel network in contact with an external solution. The hydrogel acts as semi-permeable membrane because the ions fixed on the polymer chains cannot be exchanged.

Due to the attractive power of fixed charges, the concentration of mobile ions inside the gel will exceed that in the external solution. The chemical potential difference between the external solution and the gel network can be written as

$$(\Delta\mu_{\text{s}}^{\text{ext}})_{\text{ion}} - (\Delta\mu_{\text{s}}^{\text{gel}})_{\text{ion}} = V_{\text{s,m}}RT\sum_{j}(c_j^{\text{gel}} - c_j^{\text{ext}}). \tag{2.16}$$

If one assumes the external solution to contain a concentration $c_{\text{s}}^{\text{ext}}$ of electrolyte $C_{\nu_+}^{z+}A_{\nu_-}^{z-}$ dissolved in the solvent as $\nu_+C^{z+} + \nu_-A^{z-}$, with ν_+, ν_- as the signed stoichiometric coefficients and z_+, z_- as the charge numbers, one can express the concentrations of mobile cations and anions in the external solution as

$$c_+^{\text{ext}} = \nu_+ c_{\text{s}}^{\text{ext}}, \tag{2.17}$$

$$c_-^{\text{ext}} = \nu_- c_{\text{s}}^{\text{ext}}, \tag{2.18}$$

leading to a total concentration of

$$c_+^{\mathrm{ext}} + c_-^{\mathrm{ext}} = (\nu_+ + \nu_-) \cdot c_{\mathrm{s}}^{\mathrm{ext}} = \nu c_{\mathrm{s}}^{\mathrm{ext}}. \tag{2.19}$$

Similarly, the concentrations of ions in the gel phase can be written as

$$c_+^{\mathrm{gel}} = \nu_+ c_{\mathrm{s}}^{\mathrm{gel}} \tag{2.20}$$

and

$$c_-^{\mathrm{gel}} = \nu_- c_{\mathrm{s}}^{\mathrm{gel}} + \frac{\alpha c_{\mathrm{i}}}{z_-}, \tag{2.21}$$

where $c_{\mathrm{s}}^{\mathrm{gel}}$ is the electrolyte concentration, c_{i} the concentration of ionizable groups, and α the degree of ionization. The first term in Eq. (2.21) originates from the electrolyte anions in the gel phase and the second term from the fixed ionized (negatively charged) groups on the polymer backbone. Using Eqs. (2.19)–(2.21) together with Eqs. (2.16) and (2.2) gives the ion osmotic pressure

$$\Pi_{\mathrm{ion}} = RT\left[\nu \cdot (c_{\mathrm{s}}^{\mathrm{ext}} - c_{\mathrm{s}}^{\mathrm{gel}}) - \frac{\alpha c_{\mathrm{i}}}{z_-}\right]. \tag{2.22}$$

The concentrations $c_{\mathrm{s}}^{\mathrm{ext}}$ and $c_{\mathrm{s}}^{\mathrm{gel}}$ can be related by considering that the chemical potentials of the mobile ions inside and outside the gel are equal at equilibrium ($\mu_+^{\mathrm{gel}} + \mu_-^{\mathrm{gel}} = \mu_+^{\mathrm{ext}} + \mu_-^{\mathrm{ext}}$). Therefore, the product of activities equals

$$a_+^{\mathrm{gel}\,\nu^+} a_-^{\mathrm{gel}\,\nu^-} = a_+^{\mathrm{ext}\,\nu^+} a_-^{\mathrm{ext}\,\nu^-}. \tag{2.23}$$

Assuming dilute solutions, activities and concentrations are approximately equal and, consequently, Eq. (2.23) can be rewritten as

$$\left(\frac{c_+^{\mathrm{gel}}}{c_+^{\mathrm{ext}}}\right)^{\nu_+} = \left(\frac{c_-^{\mathrm{ext}}}{c_-^{\mathrm{gel}}}\right)^{\nu_-}. \tag{2.24}$$

Eqs. (2.17)–(2.21) together with Eq. (2.24), after appropriate transposition, lead to

$$c_{\mathrm{s}}^{\mathrm{ext}} - c_{\mathrm{s}}^{\mathrm{gel}} = \frac{\alpha c_{\mathrm{i}}}{\nu z_-} - \left(\frac{1}{2z_+ z_- \nu^2}\right) \cdot \left(\frac{\alpha^2 c_{\mathrm{i}}^2}{c_{\mathrm{s}}^{\mathrm{ext}}}\right). \tag{2.25}$$

Insertion of Eq. (2.25) into Eq. (2.22) leads to the expression of the ion osmotic pressure as a function of the electrolyte concentration in the external solution as

$$\Pi_{\mathrm{ion}} = -RT\left(\frac{\alpha^2 c_{\mathrm{i}}^2}{2z_+ z_- \nu c_{\mathrm{s}}^{\mathrm{ext}}}\right). \tag{2.26}$$

Under additional consideration of the definition of ionic strength I as

$$I = \frac{1}{2}\sum_{i=1}^{n} c_i z_i^2 \tag{2.27}$$

and the assumption of a 1:1 electrolyte[1], it follows

$$\Pi_{\text{ion}} = -RT\left(\frac{\alpha^2 c_{\text{i}}^2}{4I}\right). \tag{2.28}$$

For an anionic hydrogel containing carboxyl groups (COOH) (e. g. PAA in a PVA/PAA hydrogel), the ionic equilibrium can be described by the dissociation constant

$$K_{\text{a}} = \frac{c_{\text{COO}-} \cdot c_{\text{H}+}}{c_{\text{COOH}}}. \tag{2.29}$$

The degree of ionization α can be written as

$$\alpha = \frac{c_{\text{COO}-}}{c_{\text{COOH}} + c_{\text{COO}-}} \tag{2.30}$$

and – using Eq. (2.29) – as

$$\alpha = \frac{K_{\text{a}}}{K_{\text{a}} + c_{\text{H}+}} = \frac{10^{-\text{pK}_{\text{a}}}}{10^{-\text{pK}_{\text{a}}} + 10^{-\text{pH}}}. \tag{2.31}$$

Including Eq. (2.31) in Eq. (2.28), the osmotic pressure due to ionic contribution becomes

$$\Pi_{\text{ion}} = -\frac{RT}{4I}\left(\frac{10^{-\text{pK}_{\text{a}}}}{10^{-\text{pK}_{\text{a}}} + 10^{-\text{pH}}} \cdot c_{\text{i}}\right)^2. \tag{2.32}$$

The above representation clearly reveals the dependence of the osmotic pressure on pH as well as an always existent cross-sensitivity against ionic strength when considering pH as the measuring quantity.

The total osmotic swelling pressure for an ionic hydrogel due to all three contributions reads

$$\Pi_{\text{swell}} = RT\left\{\frac{1}{V_{\text{s,m}}}\left[\ln(1-\varphi_{\text{p}}) + \varphi_{\text{p}} + \chi\varphi_{\text{p}}^2\right] + \nu_{\text{c}}\left(\varphi_{\text{p}}^{1/3} - \frac{\varphi_{\text{p}}}{2}\right) - \frac{\alpha^2 c_{\text{i}}^2}{4I}\right\}. \tag{2.33}$$

[1]Electrolyte molecules that dissociate in solution in equal numbers of anions and cations with oxidation state -1 and +1, respectively, e. g. NaCl or KCl.

2.1.2 Kinetics of the swelling process

When a sudden change in external stimulus occurs (addition of an analyte with concentration c_α to the measuring solution), the time scale for the establishment of a new equilibrium state of the hydrogel structure is crucial for the response time of hydrogel-based chemical sensors. The macroscopic swelling and shrinking of hydrogels is inevitably associated with the uptake and release of water and solvent molecules from the applied chemical stimulus. In general, hydrogel swelling kinetics is complicated due to the occurrence of simultaneous mass transfer and ionization processes. The time-dependent VPT behavior of hydrogels is mainly influenced by the kinetics of these processes and can be generally assumed as a two-step mechanism. If the chemical equilibrium between gel and solution is perturbed by a change in concentration of the external solution, a concentration gradient between both phases is created. Thus, in a first step, diffusion into or out of the gel takes place according to FICK's law. The concentration change inside the hydrogel due to diffusion of the analyte into the hydrogel can be described as [116]

$$\Delta c_\alpha^{\mathrm{gel}} = (c_\alpha^{\mathrm{ext}} - c_{\alpha,0}^{\mathrm{gel}})\left[1 - \mathrm{erf}\left(\frac{x}{2\sqrt{D_\alpha t}}\right)\right], \tag{2.34}$$

where $\Delta c_\alpha^{\mathrm{gel}}$ is the concentration change of analyte within the gel, $c_{\alpha,0}^{\mathrm{gel}}$ the initial additive concentration within the hydrogel, c_α^{ext} the analyte concentration in the external solution, $\mathrm{erf}(\delta)$ the error function, x the distance from the gel surface, D_α the effective diffusion coefficient of the analyte, and t the time. The time constant τ_α of this step can be estimated as [8]

$$\tau_\alpha = \frac{4d^2}{\pi D_\alpha}, \tag{2.35}$$

and is, thus, proportional to the square of the critical dimension d of the hydrogel layer in swelling equilibrium ($d(t \to \infty)$) and inversely proportional to the effective diffusion coefficient. In Table 2.1 the effective diffusion coefficients of selected analyte species in hydrated hydrogel networks, which directly determine the time constant of the first step, are summarized.

For the swelling of ionic hydrogels in buffered solutions, swelling kinetics of the first step is additionally governed by the properties of the buffer medium and the hydrogel, referred to as buffer-mediated diffusion reaction, as devised by LESHO and SHEPPARD [120]. The resulting time constant can be written as

$$\tau_{\mathrm{bmd}} = \frac{d^2}{\pi^2 D_{\mathrm{bm}}}\left(1 + \frac{\beta_{\mathrm{gel}}}{\beta_{\mathrm{bm}}\varphi_{\mathrm{p}}^{-1}}\right), \tag{2.36}$$

where D_{bm} is the diffusivity of the buffer medium in the hydrated gel, and β_{gel}, β_{bm}

Table 2.1: Effective diffusion coefficients of selected ions, methanol-water mixtures, and thermal diffusivity inside different water-swollen hydrogels as well as their cooperative diffusion coefficients.

Hydrogel	Diffusion coefficient in $cm^2\,s^{-1}$						Ref.
	Na^+	OH^-	H^+	MeOH in H_2O	thermal	D_{coop}	
PVA/PAA	2.9×10^{-6}	4.9×10^{-6}	8.2×10^{-6}	–	–	3×10^{-7}	[8]
PMAA	2.7×10^{-6}	–	9.4×10^{-6}	–	–		[117]
PNIPAAm	–	–	–	2.5×10^{-5}	1×10^{-3}	3×10^{-7}	[118], [119]

are the buffer capacities of the gel and the buffer medium, respectively. Thus, the time constant τ_{bmd} is influenced by the volume fraction of polymer φ_p and the ratio of the buffer capacity of gel and buffer medium.

In a second step, a macroscopic change in gel volume, i. e. a change in the degree of swelling, occurs. Due to the network expansion or contraction, the network chains adopt a new conformation. The motion of polymer chains within the solvent can be regarded as a cooperative diffusion process due to the existence of crosslinks and is characterized by a cooperative diffusion coefficient D_{coop}, first introduced by Tanaka, Hocker, and Benedek (THB-theory) by the formulation of an equation of motion [121]. The equation of motion was rewritten by Li and Tanaka as [122]

$$\frac{\partial \boldsymbol{u}}{\partial t} = D_{coop} \nabla^2 \boldsymbol{u} + \frac{K + \frac{1}{3}G}{f} \nabla \cdot (\nabla \cdot \boldsymbol{u}). \tag{2.37}$$

Here, K is the bulk modulus, G the shear modulus, f the friction coefficient between the gel network and the solvent, and $\boldsymbol{u}$ the displacement vector. The cooperative diffusion coefficient is also related to the gel properties ($D_{coop} = f(K, G, f)$). The displacement vector $\boldsymbol{u} = \boldsymbol{u}(r,t) = \boldsymbol{r}' - \boldsymbol{r}$ describes the deviation of the location $\boldsymbol{r}$ of a junction point from its average equilibrium position $\boldsymbol{r}'$ at time t. For $t \to \infty$ the displacement yields $\boldsymbol{u} = \boldsymbol{0}$. The solution of Eq. (2.37) for a spherical gel describes the time-dependent relative change of the sphere radius as [122]

$$\frac{r(\infty) - r(t)}{r(\infty) - r(0)} = \frac{6}{\pi^2} \sum_{n=1}^{\infty} n^{-2} \exp\Big(-n^2 \frac{\pi^2 D_{coop}}{r_\infty^2} t\Big) = \frac{6}{\pi^2} \sum_{n=1}^{\infty} n^{-2} \exp\Big(-n^2 \frac{t}{\tau}\Big). \tag{2.38}$$

From Eq. (2.38) the first order ($n = 1$) relaxation time constant τ_{sp} for the cooperative movement of polymer chains of a spherical gel to establish a new equilibrium conformation can be determined as

$$\tau_{\mathrm{sp}} = \frac{r_\infty^2}{\pi^2 D_{\mathrm{coop}}}. \tag{2.39}$$

The higher order relaxation time constants ($n > 1$) can be neglected for long times. Similarly to the time constant for the diffusion of solvent molecules in Eq. (2.35), this time constant is proportional to the square of the critical gel dimension, i. e. the radius r of the gel for $t \to \infty$, and inversely proportional to the cooperative diffusion coefficient D_{coop}. The cooperative diffusion coefficient has been determined by different groups to be in the order of $10^{-7}\,\mathrm{cm^2/s}$ for various hydrogels [8, 121]. It is approximately one to two orders of magnitude smaller than the diffusion coefficient D_α of analyte molecules as can be seen from Table 2.1. Hence, the time-dependent VPT and, consequently, the sensor response time are primarily determined by the time scale of the motion of the polymer chains in solution.

According to Li and Tanaka [122], the cooperative swelling process is a two-step process. The first step is a pure diffusion process followed by a shear process which minimizes the anisotropic deformation. For symmetric hydrogel geometries like spheres, the energy loss due to shear is minimal (shear modulus $G = 0$). In contrast, for other gel geometries, e. g. slabs which may be even more important for sensor applications, the contribution of the shear relaxation process has to be considered ($G \neq 0$) leading to a reduced speed of the cooperative diffusion process. Thus, an additional shape factor f_{s}, which reduces the effective diffusion coefficient compared to a symmetric gel shape, as well as a correction term a have to be considered. The time constant due to cooperative diffusion can then be written in a more general form as

$$\tau = \frac{d_{\mathrm{c}}^2}{f_{\mathrm{s}} a^2 D_{\mathrm{coop}}}. \tag{2.40}$$

The correction term a is the root of the first order Bessel function and also depends on the ratio R of shear and bulk modulus (Fig. 2.2a), where R can be written as

$$R = \frac{\frac{G}{K}}{1 + \frac{4}{3}\frac{G}{K}}. \tag{2.41}$$

The linear critical dimension d_{c}, the shape factor f_{s}, the correction term a, and the effective cooperative diffusion coefficient $D_{\mathrm{coop,eff}}$ for selected geometries are listed in Table 2.2.

Because of the non-zero shear modulus for unsymmetrical gel geometries, the directions in which no diffusion takes place are subjected to the shear relaxation process, reducing the speed of the cooperative diffusion process in the diffusion direction. Therefore, this reduction is most pronounced for disc-like hydrogel structures, because here diffusion

Table 2.2: Parameters influencing the time constant of the VPT of hydrogels – due to cooperative diffusion – for selected gel geometries. The given range for the lower and upper limit for the correction term a corresponds to $G/K = 0$ and $G/K \to \infty$, respectively [122].

Gel geometry	d_c	f_s	a	$D_{coop,eff}$
Sphere	Radius (r_∞)	1	$\pi \ldots 0$	D_{coop}
Cylinder	Radius (r_∞)	2/3	2.74 … 0	$\frac{2}{3}D_{coop}$
Disc	Thickness (d_∞)	1/3	2.29 … 0	$\frac{1}{3}D_{coop}$

only occurs in uniaxial direction and the two remaining directions are subjected to shear relaxation. The time constant for initial analyte diffusion into the hydrogel (first step) and the time constants for cooperative movement of the hydrogel network (second step) as a function of the critical linear hydrogel dimension d_c according to Eqs. (2.35) and (2.40) are shown in Fig. 2.2 for a PVA/PAA hydrogel and selected gel geometries. From Fig. 2.2b it becomes clear that cooperative diffusion is the dominant effect for the time constant of macroscopic hydrogel swelling which is greatly influenced by the geometric properties of the hydrogel. A reduction of the critical linear dimension d_c will lead to smaller response times of the network. However, even for the sub-100 µm range, response times in the order of tenth of seconds can be expected. The model is in excellent agreement with experimental data for unconfined free swelling hydrogels and depicts the important parameters influencing swelling

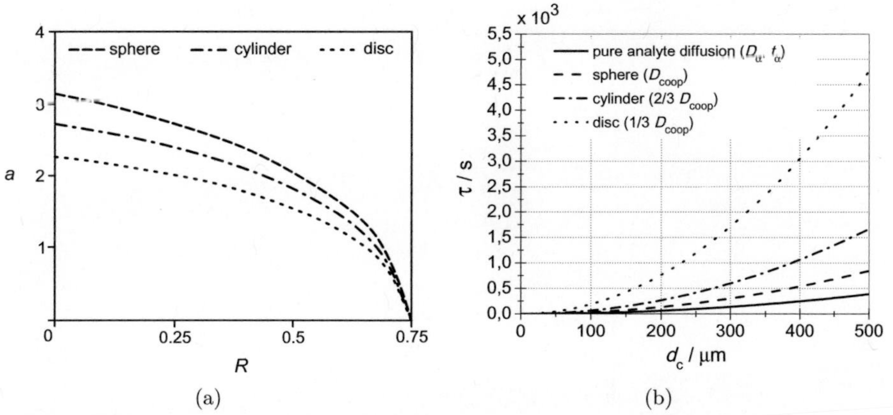

(a) (b)

Figure 2.2: (a) Correction term a as a function of R, where $R = 0$ and $R = 0.75$ corresponds to $G/K = 0$ and $G/K \to \infty$, respectively (reproduced from [123]), (b) comparison of the time constants for the VPT of a PVA/PAA hydrogel for pure solvent diffusion (diffusion of H^+ considered, solid line) and cooperative diffusion for selected gel geometries and $R = 0$ (interrupted lines) as a function of the critical linear dimension of the gel.

kinetics. However, for application of hydrogels in microsensors or MEMS, swelling kinetics might be more complicated and the prevailing mechanical constraints, which might further reduce the speed of swelling, have to be considered.

2.1.3 Linear mechanical hydrogel models

The macroscopic volume change of polymer gels can be adequately described by separating the deformation problem in two parts and subsequent superposition of both parts. One part describes the elastic properties of the gel network and the second part describes the behavior of the interstitial fluid. The rubber-like elastic behavior is characterized by an entropy change and can be modeled by a linear spring element with YOUNG's modulus E (equivalent to spring stiffness) according to HOOKE's law. The viscous flow of a fluid under shear deformation can be described by a dashpot of viscosity η according to NEWTON's law. The combination of these basic linear mechanical elements enables different models for the description of the viscoelastic behavior of polymeric hydrogels. A systematic combination of these elements to two- or multi-parameter models allows linear modeling of specific mechanical aspects, like drift or relaxation, accompanying the VPT. Selected mechanical models and the differential equations describing them, are shown in Table 2.3.

Table 2.3: Selected linear lumped models for the description of the viscoelastic material properties of polymeric hydrogels [124] where σ and ε are the mechanical stress and strain, respectively.

Model type	Network structure	Differential equation
MAXWELL		$\dot{\sigma} + \frac{E}{\eta}\sigma = E\dot{\varepsilon}$
VOIGT-KELVIN	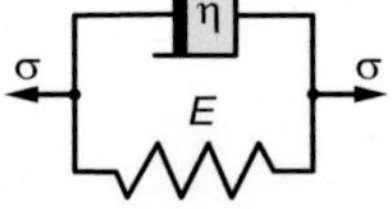	$\sigma = E\varepsilon + \eta\dot{\varepsilon}$
ZENER	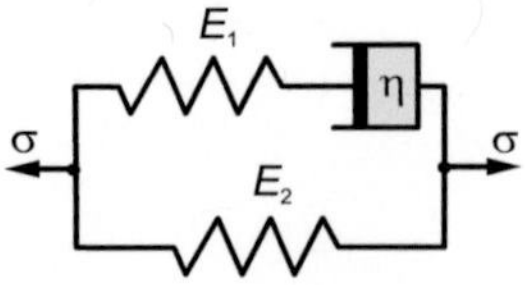	$\dot{\sigma} + \frac{E_1}{\eta}\sigma = (E_1 + E_2) \cdot \dot{\varepsilon} + \frac{E_1 E_2}{\eta}\varepsilon$
BURGER	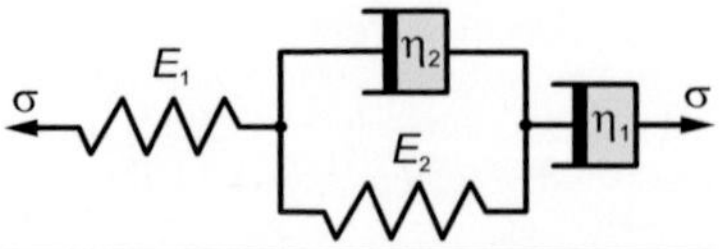	$\ddot{\sigma} + (\frac{E_1}{\eta_1} + \frac{E_1+E_2}{\eta_2}) \cdot \dot{\sigma} + \frac{E_1 E_2}{\eta_1 \eta_2}\sigma =$ $E_1 \cdot \ddot{\varepsilon} + \frac{E_1 E_2}{\eta_2}\dot{\varepsilon}$

2.2 Piezoresistive pressure transducers

2.2.1 Piezoresistive effect

Piezoresistive pressure sensors are monolithic microdevices which make use of the piezoresistive effect in semiconductors. These devices are able to convert a mechanical force or a pressure into an electrical output signal. The majority of applications is fabricated by micromachining techniques using semiconducting monocrystalline silicon Here, the geometric peculiarities due to the regular crystalline structure as well as the possibility of adaptable electro-physical properties are advantageously combined. The piezoresistive effect describes a mechano-electrical effect where the resistivity and, consequently, the resistance R of a material is changed due to applied mechanical load (stress, strain). The resistance change ΔR of such a piezoresistor under the assumption of a small length and cross-section change (usually much less then 1 %) can be written as

$$r = \frac{\Delta R}{R} = \left[2 - \frac{\Delta(N\mu)}{\varepsilon N\mu}\right]\varepsilon = K\varepsilon \tag{2.42}$$

and depends on the number of charge carriers N, their mobility μ, and the mechanical strain ε [125]. For semiconductor materials, the second term in brackets of Eq. (2.42) is most pronounced compared to the sole geometric effect (first term in brackets, i. e. factor 2), because the number of charge carriers and their mobility significantly change with deformation due to the deformed band structure of the material. This directly reflects in the K-factor which is about $K = 2$ for metal thin films and can be up to 100 for semiconductors. For piezoresistive pressure transducers, those piezoresistors are integrated into a silicon deformation body, i. e. a flexure plate, next to the surface. At the interface between piezoresistor and silicon, the mechanical stress and strain of both is equal. Because of the crystalline structure of semiconductors, the K-factor shows anisotropic characteristics. This behavior is described by piezoresistive coefficients π_{ij} which are components of a 6×6 matrix $\mathbf{\Pi}$ (Voigt-notation for a cubic lattice structure) relating the relative resistance change to the mechanical stress (vector) $\boldsymbol{\sigma}$ as

$$\boldsymbol{r} = \mathbf{\Pi} \cdot \boldsymbol{\sigma}. \tag{2.43}$$

In case of silicon, the number of independent piezoresistive coefficients is reduced to three, namely π_{11}, π_{12}, and π_{44}. Thus, these coefficients directly determine the resistance change for a constantly applied stress and depend on the semiconductor type (n-type or p-type), the temperature, the doping concentration, as well as on the orientation of the resistor structure in the lattice of the deformation body. The

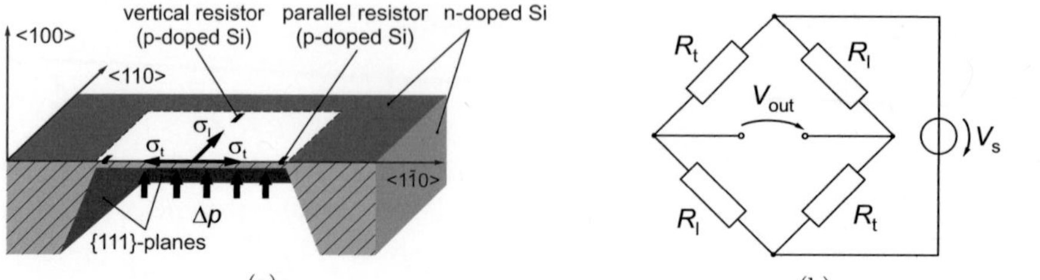

(a) (b)

Figure 2.3: (a) Silicon pressure sensor chip with a square flexure plate acting as deformation body for vertically and parallel aligned p-silicon piezoresistors in <110>-direction; (b) WHEATSTONE full bridge configuration of parallel ($R_t = R_0(1 + r_t)$) and perpendicularly ($R_l = R_0(1 + r_l)$) aligned piezoresistors with bridge voltage V_{out} and supply voltage V_s.

highest value of the piezoresistive coefficients can be found to be π_{44} for p-silicon ($\pi_{44} = 138.1 \times 10^{-11}\,\mathrm{Pa}^{-1}$ at r. t.). A maximum utilization of this coefficient is achieved if the resistor structure is located in the (100)-plane of the deformation body and rotated by 45° to be oriented in <110>-direction (Fig. 2.3a). For this case, the longitudinal and transversal piezoresistive coefficients can be written as [125]

$$\pi_l = \frac{1}{2}(\pi_{11} + \pi_{12} + \pi_{44}) \qquad \pi_t = \frac{1}{2}(\pi_{11} + \pi_{12} - \pi_{44}), \tag{2.44}$$

and simplified to

$$\pi_l \approx -\pi_t \approx \frac{\pi_{44}}{2} \tag{2.45}$$

for p-silicon, because π_{11} and π_{12} are only 4.8 % and -0.8 % of π_{44}, respectively. π_l and π_p denominate the longitudinal and transverse piezoresistive effect, i. e. whether the applied mechanical stress acts parallel or vertical to the direction of the current inside the piezoresistors.

For resistor structures in a biaxial stress field, the resistance change is often expressed by a longitudinal and a transverse component as

$$r_l = \pi_l\sigma_l + \pi_t\sigma_t, \qquad r_t = \pi_t\sigma_l + \pi_l\sigma_t. \tag{2.46}$$

The location of two piezoresistors parallel and two resistors perpendicular to the edge of the flexure plate (Fig. 2.3a) leads to transversal strain of the parallel resistors and longitudinal strain of the perpendicular resistors. Due to the different signs of the longitudinal and transversal piezoresistive coefficients as stated in Eq. (2.45), they can be combined in a full bridge circuit (Fig. 2.3b). The bridge voltage V_{out} in relation to the fixed supply voltage can then be written as [125]

$$\frac{V_{\mathrm{out}}}{V_{\mathrm{s}}} = \frac{r_{\mathrm{t}} - r_{\mathrm{l}}}{2 + r_{\mathrm{t}} + r_{\mathrm{l}}} \approx \frac{1}{2}(r_{\mathrm{t}} - r_{\mathrm{l}}) = \frac{1}{2}(\pi_{\mathrm{l}} - \pi_{\mathrm{t}})(\sigma_{\mathrm{l}} - \sigma_{\mathrm{t}}) = \frac{1}{2}\pi_{44}(\sigma_{\mathrm{l}} - \sigma_{\mathrm{t}}). \qquad (2.47)$$

In case of the application of a constant current supply, the temperature coefficient of the sensitivity (TCS) depends on the temperature coefficient of the specific resistance (TCR) and the temperature coefficient of the piezoresistive coefficients ($TC\pi$). Because of the different signs of TCR and $TC\pi$, two specific doping concentrations for the piezoresistors can be found to compensate both temperature coefficients ($TCR = -TC\pi$) [102]. From Eq. (2.47) it becomes clear that the accomplishment of a high transducer sensitivity directly implies a maximization of the resulting stress at the location of the piezoresistors. In other words, the piezoresistors have to be placed where the maximum stress occurs , which in turn depends on the geometric properties of the deformation body.

2.2.2 Influence of the geometric properties of the flexure plate on sensitivity

The deformation w of a flexure plate can be described by a partial differential equation (PDE) of order four. Under the assumption of a steady displacement under uniform pressure, the PDE can generally be written as [126]

$$D\left(\frac{\partial^4 w}{\partial x^4} + 2\frac{\partial^4 w}{\partial x^2 \partial y^2} + \frac{\partial^4 w}{\partial y^4}\right) = p, \qquad (2.48)$$

with flexural rigidity

$$D = \frac{Eh^3}{12(1-\nu^2)}. \qquad (2.49)$$

Here, E is YOUNG's modulus, h the thickness of the flexure plate, and ν POISSON's ratio. The solution of Eq. (2.48) for the boundary condition that all edges are clamped can be found in Table 2.4 for rectangular and circular plate geometries. The maximum stress as well as the transducer sensitivity for maximal stress are derived from the deformation function and are also stated in Table 2.4.

From a technological point of view, the thickness and the geometric dimensions of the flexure plate are the only parameters which can be altered in order to attain high sensitivity. A reduction of the plate thickness leads to an increase in sensitivity (Fig. 2.4a) of more than one order of magnitude. However, for a given pressure, a thinner plate results in a larger deflection accompanied by an extension of the neutral axis which leads to a larger nonlinearity because the membrane stress is proportional to even powers of p. Especially for the small pressure range well below 100 kPa, the overall nonlinearity is determined by the membrane stress. In this case,

Table 2.4: Selected mechanical quantities of rectangular and circular flexure plates as deformation body of piezoresistive transducers as well as the corresponding transducer sensitivities. The solution $w(x, y)$ of Eq. (2.48) for the rectangular plate is based on an energy minimization approach, whereas the solution $w(r)$ for the circular plate is the exact analytical solution according to [127]. $z = h/2$ describes the upper surface of the flexure plate.

Plate geometry	Rectangular	Circular
Schematic	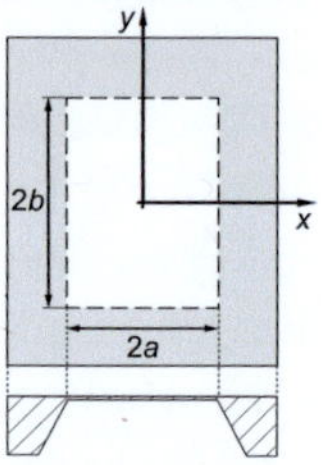	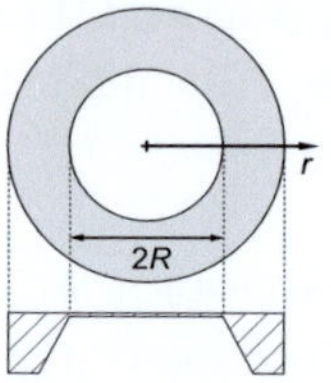
Deflection	$w(x,y) = w_0(1-\frac{x^2}{a^2})^2(1-\frac{y^2}{b^2})^2$ $w_0 = \frac{7a^4b^4}{128D(a^4+b^4+\frac{4}{7}a^2b^2)}p$	$w(r) = w_0\left(1-\frac{r^2}{R^2}\right)^2$ $w_0 = \frac{pa^4}{64D}$
Maximum stress	$\sigma_{\mathrm{x,max}} = \sigma(x=a, y=0, z=h/2)$ $= -\frac{21a^2b^4}{8h^2(a^4+b^4+\frac{4}{7}a^2b^2)}p$ $\sigma_{\mathrm{y,max}} = \nu \cdot \sigma_{\mathrm{x,max}}$	$\sigma_{\mathrm{r,max}} = \sigma(r=R, z=h/2)$ $= \frac{3R^2}{4h^2}p$ $\sigma_{\mathrm{tan,max}} = \nu \cdot \sigma_{\mathrm{r,max}}$
Sensitivity	$S = \frac{V_{\mathrm{out}}/V_{\mathrm{s}}}{p} = -\frac{21\pi_{44}(1-\nu)a^2b^4}{16h^2(a^4+b^4+\frac{4}{7}a^2b^2)}$	$S = \frac{V_{\mathrm{out}}/V_{\mathrm{s}}}{p} = \frac{3\pi_{44}(1-\nu)R^2}{8h^2}$

assuming a maximum acceptable nonlinearity, a reduction of the plate thickness to enhance the sensitivity is only possible in a limited range. From Fig. 2.4b one can observe the sensitivity of a rectangular plate normalized to the sensitivity of a circular plate as a function of the aspect ratio of the rectangular plate. This figure reveals a dependence of the plate geometry on the transducer sensitivity and clearly shows that a square bending plate will show the highest sensitivity (about 7 % higher compared to a circular plate with the same surface area). Increasing the aspect ratio of a rectangular plate leads to a decrease in sensitivity compared to a circular plate. This favours the application of piezoresistive transducers with a square flexure plate as mechano-electrical transduction elements for hydrogel-based sensors.

2.2.3 Influence of local stiffening of the flexure plate

Flexure plates with a stiffened center, as shown in Fig. 2.5, are characterized by a localized rigidity increase, resulting in altered mechanical properties of the flexure plate under pressure. The mechanical energy stored in the flexure plate is concentrated

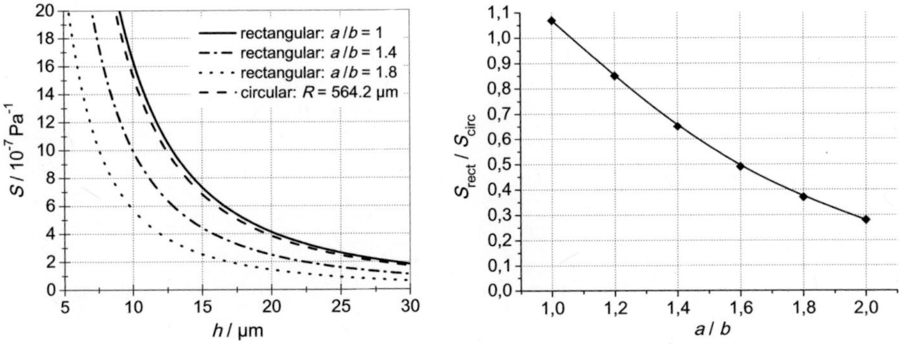

(a) (b)

Figure 2.4: Sensitivity of silicon piezoresistive pressure transducers: (a) for circular and rectangular plates with different aspect ratios as a function of the plate thickness, (b) of a rectangular flexure plate normalized to the sensitivity of a circular plate with the same surface area. A surface area of 1 mm^2 has been considered for the calculations.

at the end of the rigid center ($x = R_{\mathrm{B}}$) and at the edge of the flexure plate ($x = R$). These structures are used as pressure sensors for small pressures but are also suitable as actuator components because the stiffened center part, also referred to as boss, can move in vertical direction.

To point out the characteristics of a stiffened flexure plate, for the sake of simplicity, a circular plate geometry, governed by relatively simple equations, is assumed. However, the conclusions are qualitatively adaptable to rectangular or square plate geometries. Under the assumption that the thickness of the boss is at least six times larger than the plate thickness and that $R_{\mathrm{B}}/R > 0.15$, the relation between the applied pressure and the corresponding static deflection can be described as [128]

$$p = \frac{hE}{R^4} w_0 \left(A_{\mathrm{p}} \frac{h^2}{1-\nu^2} + B_{\mathrm{p}} w_0^2 \right), \tag{2.50}$$

where A_{p} and B_{p} denote stiffness coefficients for the linear and the non-linear term

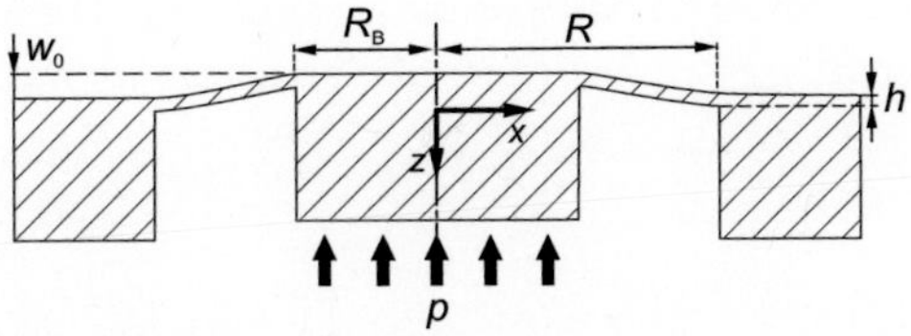

Figure 2.5: Cross-sectional illustration of a clamped flexure plate with central stiffening (boss structure).

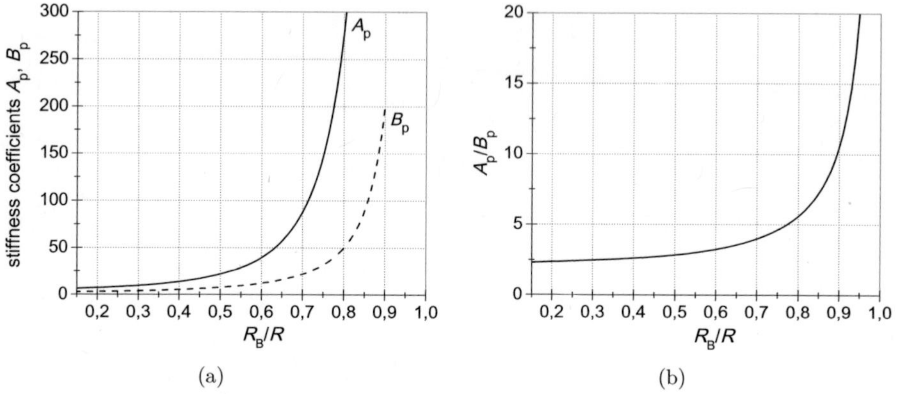

Figure 2.6: (a) Linear and non-linear stiffness coefficients as function of the radius of the boss structure, according to Eqs. (2.51) and (2.52), (b) ratio of the stiffness coefficients as a function of the radius of the boss structure [128]. A POISSON's ratio of $\nu = 0.063$ has been used.

in Eq. (2.50), respectively. The linear stiffness coefficient A_p only depends on the lateral dimensions of the boss structure and the flexure plate, respectively, as

$$A_\mathrm{p} = \frac{16}{3 - 3\left(\frac{R_\mathrm{B}}{R}\right)^4 - 12\left(\frac{R_\mathrm{B}}{R}\right)^2 \ln\left(\frac{R}{R_\mathrm{B}}\right)}, \tag{2.51}$$

whereas the non-linear coefficient B_p also depends on the material properties as

$$B_\mathrm{p} = \frac{\frac{(7-\nu)}{3}\left[1 + \left(\frac{R_\mathrm{B}}{R}\right)^2 + \left(\frac{R_\mathrm{B}}{R}\right)^4\right] + \frac{(3-\nu)^2}{(1+\nu)}\left(\frac{R_\mathrm{B}}{R}\right)^2}{(1-\nu)\left[1 - \left(\frac{R_\mathrm{B}}{R}\right)^4\right]\left[1 - \left(\frac{R_\mathrm{B}}{R}\right)^2\right]}. \tag{2.52}$$

From Fig. 2.6 an increase of both coefficients as well as their ratio can be observed, if the radius of the boss structure is increased. With increasing $A_\mathrm{p}/B_\mathrm{p}$, A_p always exceeds B_p resulting in an increasing suppression of the non-linear term. Thus, the larger the stiffened region is, the smaller is the remaining non-linearity. On the other hand, the larger the coefficients are, the smaller is the deflection at constant pressure, which is related to a decrease in sensitivity. Furthermore, the large coefficients lead to the fact that an additional term due to residual stress, e. g. tensions due to packaging which would apply additively to Eq. (2.50), becomes less dominant.

II Hybrid hydrogels as sensor material

3 Sensors based on porous hydrogels [1]

This chapter addresses an approach were porous p(AAm-*co*-AA)-hydrogels, i. e. hydrogels whose morphology is altered by an open, interconnected channel structure, are studied with respect to their application as pH-sensing material in piezoresistive microsensors to improve the sensor response time. In particular, the porosity is introduced by a porogen-leaching method using different pore-forming agents. The resulting diverse hydrogel structures are characterized with respect to their response time by free-swelling experiments in order to determine the pore-forming agent which works best. The hydrogel which shows the fastest swelling kinetics is further characterized by different analysis techniques and finally integrated as sensing layer into piezoresistive chemical microsensors. The sensor behavior is shown and the sensor properties are discussed.

3.1 Porous hydrogels

3.1.1 Classification and features

Porous hydrogels can be described to consist of a polymer-rich phase (matrix material) and a solvent-rich phase (pores). According to the International Union of Pure and Applied Chemistry (IUPAC), they can be classified with respect to the pore size as [131]

- microporous ($< 2\,\mathrm{nm}$),
- mesoporous ($2\,\mathrm{nm} \ldots 50\,\mathrm{nm}$),
- macroporous ($50\,\mathrm{nm} \ldots 1\,\mu\mathrm{m}$).

However, this definition is not always strictly considered in literature and gels with pores larger than $1\,\mu\mathrm{m}$ are sometimes called macroporous or superporous [132]. In principle, all hydrogels have pores, i. e. spacing between polymer chains, also known as "molecular porosity". However, these pore structures are non-permanent, i. e. in contrast to porous hydrogels, a homogeneous hydrogel becomes non-porous when the interstitial fluid is removed.

[1] This chapter is the result of a close cooperation with the Leibniz Institute of Polymer Research Dresden e. V. (IPF), Institute for Macromolecular Chemistry. Especially hydrogel synthesis was carried out in cooperation with Dr. S. Zschoche and H. P. Zhang [129, 130].

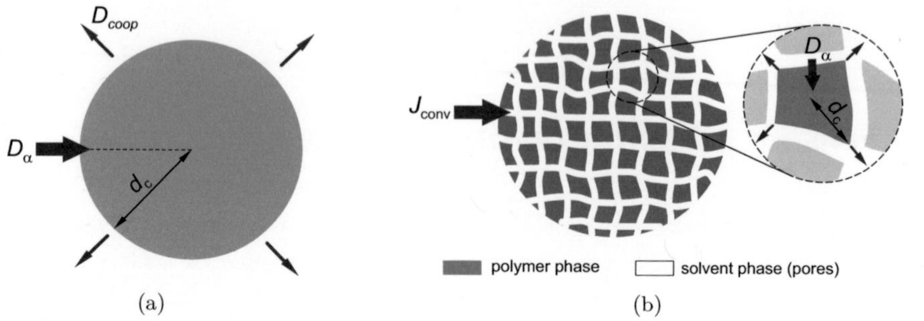

Figure 3.1: Swelling (a) of conventional non-porous hydrogels, governed by analyte diffusion (D_α) and cooperative diffusion (D_{coop}), respectively, both determined by the critical hydrogel dimension d_c, (b) of porous hydrogels where small polymer sub-units are separated by interconnected pores. Solution transport into the gel takes place by a convective flow J_{conv} and the critical dimension d_c is the half length between pores (For simplicity and explanatory reasons, a regular pore structure is shown which might not be the case for most porous hydrogels).

From an application point of view, such a heterogeneous network structure is believed to be advantageous with respect to sensor applications because:

- fast convective solution transport J_{conv} through the pores into the gel is possible due to capillary effects (Fig. 3.1b), compared to slow analyte diffusion in homogeneous gels (Fig. 3.1a);
- a large surface-to-volume ratio is created which enables an enhanced contact area of the solution with the gel;
- small polymer sub-units seperated by the pores are created. Thus, the critical dimension d_c, which is directly related to the macroscopic size of the gel body for homogeneous gels and is several 10 µm for hydrogel layers in piezoresistive microsensors, can be greatly reduced to several µm (Fig. 3.1b). This will directly lead to a decreasing time constant due to cooperative diffusion ($\tau \propto d_\mathrm{c}^2$, cp. Section 2.1.2).

3.1.2 Preparation methods

Phase separation is known as the underlying phenomenon for the formation of the porous structure, i. e. polymerization takes place in the phase-separated state [133]. Thus, different methods which induce a phase separation can be used to fabricate porous hydrogels. Here, pore size, porosity, and pore size distribution depend on the particular preparation method which includes solvent-casting/particle-leaching, gas-foaming, freeze-drying (lyophilization), freeze-thawing, and electrospinning [134].

Especially the first method can be simply integrated into the conventional synthesization process without the need for additional sophisticated equipment. Here, a porogen, which can be a solid material or a liquid, is added to the pre-gel solution. During the polymerization reaction, the porogen causes a phase separation into a polymer-rich phase (hydrogel matrix) and a polymer-poor phase (porogen material). After polymerization, the porogen is leached out from the hydrogel matrix by an appropriate selective solvent resulting in a porous network structure. This method is especially suitable for high-throughput manufacturing of hydrogel sensing layers. Furthermore, a great variety of porogens exists which enables great flexibility towards the fabrication of porous gels with targeted properties such that this method has been used within this work.

3.1.3 Hydrogel synthesis

For fabrication of porous p(AAm-*co*-AA) hydrogels, AA and AAm monomers were mixed with the porogen, *N,N,N',N'*-tetramethylethylenediamine (TEMED) as catalyst, *N,N'*-methylenebisacrylamide (BIS) as crosslinker, and sodium peroxodisulfate (NaPS) as initiator. To study the influence of different porogens on the hydrogel properties, synthesis was carried out in the presence of different organic and inorganic porogens P_n, namely NaCl, $CaCO_3$, 1,4-dioxane, and poly(ethylene glycol)

Table 3.1: Porogens used for the synthesis of porous hydrogels, their weight ratio and corresponding sample codes.

P_n	Porogen	Porogen content in wt% *	Sample code	Appearance
P_0	none	–	NP	transparent
P_1	NaCl	6.1	NaCl-6	transparent
		18.4	NaCl-18	transparent
		24.5	NaCl-24	transparent
P_2	$CaCO_3$	10	CaCO3-10	milky
		100	CaCO3-100	white (impure)
P_3	1,4-dioxane	–	Diox-10	white
P_4	PEG1000	32	PEG1-32	white
	PEG6000	32	PEG6-32	white
	PEG6000	50	PEG6-50	white
	PEG6000	80	PEG6-80	white
	PEG8000	32	PEG8-32	white

* relative to total monomer mass (cp. Appendix A.1.1).

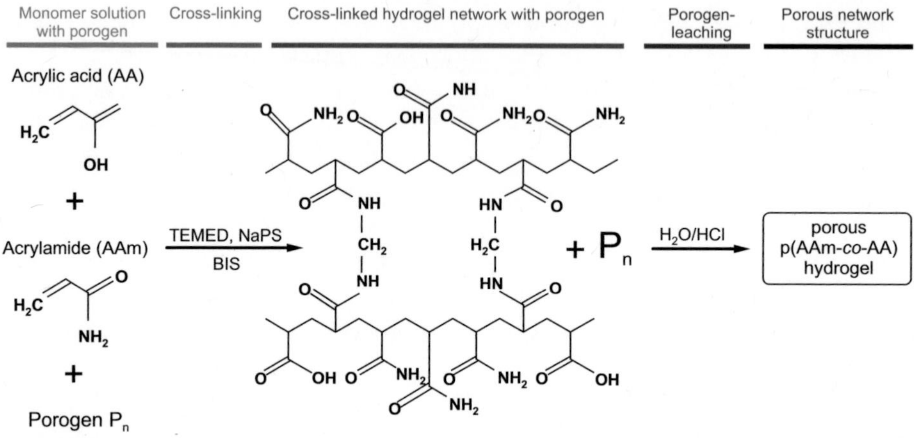

Figure 3.2: Synthesis procedure of porous p(AAm-*co*-AA) hydrogels by porogen leaching using different porogens P_n.

(PEG) with different molecular weight ($M_n = 1000\,\mathrm{g\,mol^{-1}}$ (PEG1000), $6000\,\mathrm{g\,mol^{-1}}$ (PEG6000), $8000\,\mathrm{g\,mol^{-1}}$ (PEG8000) (Fig. 3.2, Table 3.1). For comparison, also homogeneous, non-porous hydrogels have been fabricated using the same procedure, but without using porogens. Polymerization was carried out in cylindrical glass vessels with a diameter of 3.5 mm and 1 mm, respectively to obtain exactly defined hydrogel structures for characterization (see Section 3.2). After the synthesis process, the porogen has been leached out from the cross-linked hydrogel network by washing in water (for NaCl, PEG, 1,4-dioxane) or hydrochloric acid (for $CaCO_3$). A detailed description of the synthesis procedure and the chemicals used can be found in Appendix A.1.1.

3.2 Hydrogel characterization

To study the hydrogel's swelling behavior with particular emphasis on response time, time-dependent swelling curves are recorded for the fabricated cylindrical hydrogels by free-swelling experiments using a gravimetric method. The response time for the porous hydrogels is directly compared to that of non-porous hydrogels (NP). Based on the obatained results, the hydrogel with the fastest swelling response is studied in detail concerning removal of the porogen from the network, network morphology as well as pore size and pore size distribution.

3.2.1 Free-swelling measurements

For free-swelling experiments, the mass swelling degree Q_m with

$$Q_\mathrm{m} = \frac{m_\mathrm{p} + m_\mathrm{s}}{m_\mathrm{p}}. \tag{3.1}$$

is used to determine the time-dependent swelling behavior of different hydrogel samples for step changes of the pH value. Here, m_p and m_s are the mass of polymer and solvent, respectively.

a. NaCl as porogen

The time-dependent pH response for hydrogels synthesized with different NaCl concentrations are shown in Figs. 3.3a,b. Here, the response times of non-porous hydrogels and hydrogels synthesized with different NaCl content are almost identical and no response time reduction can be observed for NaCl contents up to 24.5 wt%. The

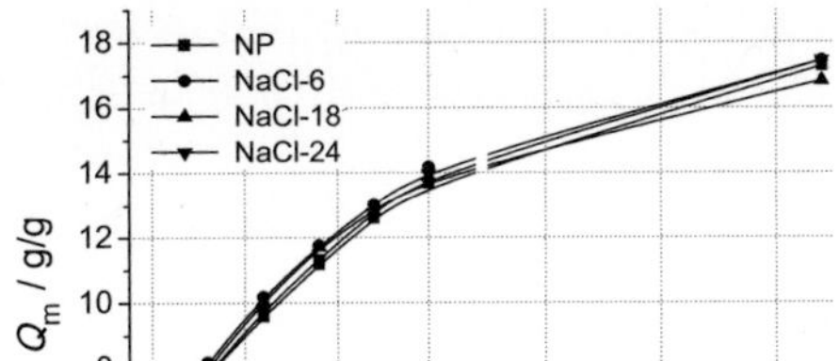

(a)

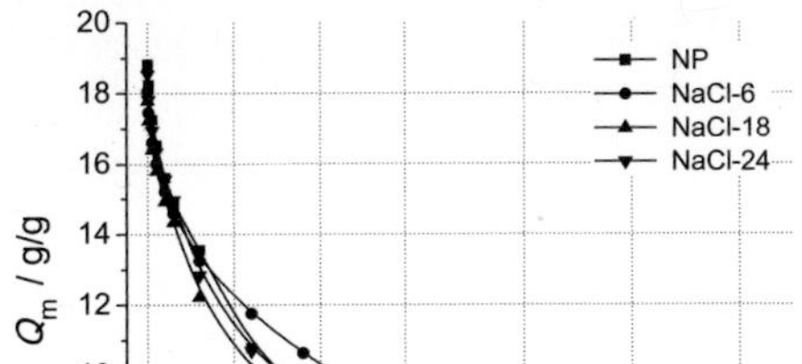

(b)

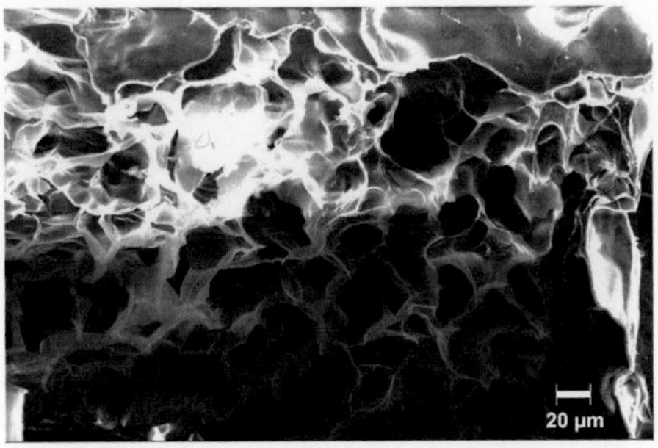

(c)

Figure 3.3: Characteristics of p(AAm-*co*-AA) hydrogels synthesized with NaCl as porogen: (a) time-dependent mass swelling degree for a pH change between pH 2 and pH 10, (b) time-dependent mass swelling degree for a pH change between pH 12 and pH 4 (sample diameter 3.5 mm), (c) SEM image of a freeze-dried NaCl-24 hydrogel.

time to reach equilibrium was found to be more than 24 h for all hydrogels. Scanning electron microscopy (SEM) studies of NaCl-24 hydrogels reveal the formation of a porous structure (Fig. 3.3c) with a pore size in the range of 20 µm. From these measurements, it can be assumed that only single closed pores do exist and the formation of an interconnected channel network is not achieved or that the observed pores are only at the surface and were caused by freeze drying. The assumption that no porous network is formed is also confirmed by the transparency of the samples which indicates that no phase separation has occurred. Furthermore, Fig. 3.3c reveals very thin polymer walls between pores thus that the collapse of these walls during swelling and the resulting formation of a dense polymer surface, which hinders the solution transport, might be possible.

b. $CaCO_3$ and 1,4-dioxane as porogens

For the preparation of porous hydrogels, $CaCO_3$ particles can be used as foaming agent and as porogen, respectively. In the first case, CO_2 is generated prior to network formation by the reaction of $CaCO_3$ with the carboxyl groups of AA as

$$CaCO_3 + 2RCOOH \rightarrow 2RCOO^- + Ca^{2+} + CO_2 + H_2O. \tag{3.2}$$

The CO_2 gas formation leads to a "foaming" of the hydrogel. The resulting pore diameter can be estimated by the YOUNG-LAPLACE equation and is in the range of 100 µm . . . 200 µm [135]. For the application of hydrogels in piezoresistive microsensors, a layer thickness between 50 µm . . . 100 µm is aimed such that this method cannot be used for the targeted application. To use $CaCO_3$ as a porogen, the –COOH groups have to be neutralized within the monomer solution using NaOH. To form a porous network structure, $CaCO_3$ is removed by HCl after polymerization. The generated

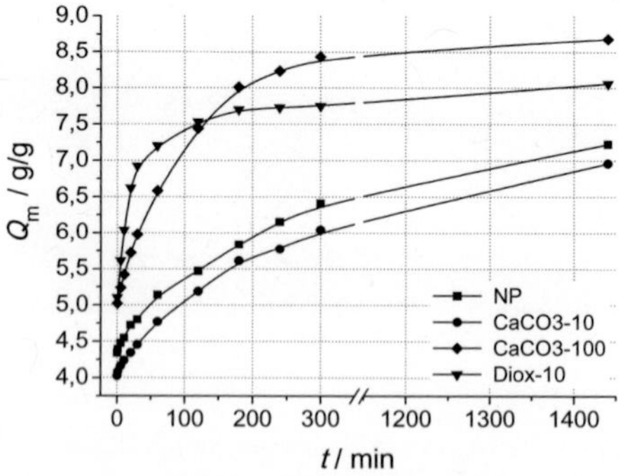

Figure 3.4: Time-dependent mass degree of swelling for a pH step between pH 2 and pH 4 for hydrogels fabricated with $CaCO_3$ and 1,4-dioxane as porogen.

CO_2 during this reaction can escape through the pores which are already formed. A direct comparison of the time-dependent mass swelling degree of hydrogels synthesized in the presence of $CaCO_3$ and 1,4-dioxane reveal that a low $CaCO_3$ content has no effect on the response time and that the response time is comparable to that of non-porous gels (Fig. 3.4). However, a tenfold increase of the $CaCO_3$ content shows a significant enhancement of the response time. The samples where 1,4-dioxane was used for pore formation show similar characteristics but with an even faster initial swelling phase. For these gels, equilibrium is almost reached after 5 h but, nevertheless, a slow drift-like re-swelling can be observed for $t > 5$ h. The hydrogel samples which show an improved response time also show a higher swelling degree at $t = 0$, i. e. a higher solution uptake due to the porous network structure.

Even though a smaller response time could be observed for the CaCO3-100 hydrogels, this gel showed no viscous rubber-like properties as it is typical for hydrogels but was comparable to a solid. This indicates that a high volume fraction of $CaCO_3$ particles might determining the mechanical properties of the gel.

c. PEG as porogen

PEG (Fig. 3.5) is a linear or branched neutral polyether that is water-soluble and available with average molecular weights between 200 g mol^{-1} ... 35000 g mol^{-1}. It is non-toxic and is, thus, often used for biomedical and cell-biological applications. In aqueous environments, this polymer has the ability of excluding other polymers from its vicinity and is thus especially suitable for the formation of a two-phase system [136].
The comparison of hydrogels synthesized in the presence of PEG with different molecular weight and constant weight ratio indicates that the use of PEG1000 is not advantageous with respect to a response time reduction (Fig. 3.6a). Here, a similar time-dependent swelling characteristics as for non-porous gels was found. However, hydrogels which were fabricated using PEG with higher molecular weights (PEG6000, PEG8000) show a significant response time reduction with a time to reach equilibrium of about 5 h. From Fig. 3.6a it can be seen that PEG6-32 shows the fastest swelling behavior.

The variation of the weight ratio of PEG during polymerization between 32 wt% and 80 wt% appears to have no influence on the response time (Fig. 3.6b). This

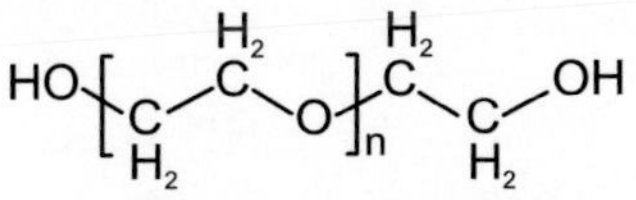

Figure 3.5: Chemical structure of a PEG macromolecule [136].

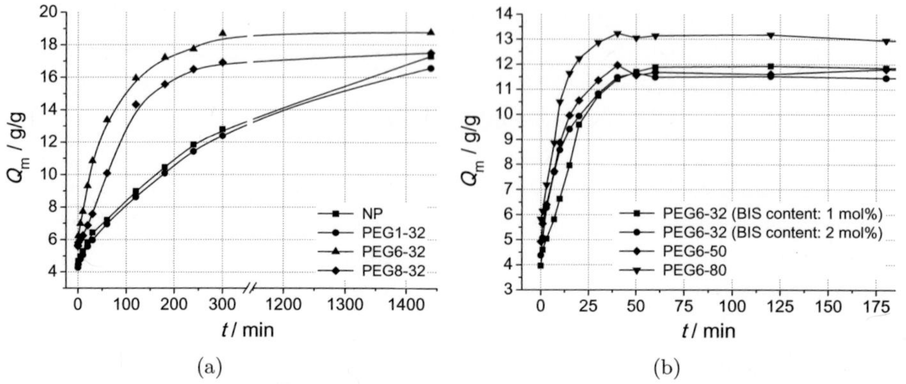

Figure 3.6: Time-dependent mass degree of swelling for a pH step between pH 2 and pH 7 for hydrogels fabricated with PEG: (a) 32 wt% PEG with varying molecular weight (sample diameter 3.5 mm), (b) PEG-6000 with varying PEG content and crosslinker concentration, respectively (sample diameter 1 mm).

suggests that even for a low PEG content of 32 wt%, the percolation threshold is already reached and an interconnected channel structure is successfully established. The variation of the crosslinker concentration for PEG6-32 hydrogels has been found to have only marginal influence on the response time and the equilibrium degree of swelling and is thus not been considered any further regarding response time reduction (Fig. 3.6b).

To study if the process of fast swelling is reversible and also occurs after shrinking and reswelling, Q_m has been measured for cyclic pH changes between pH 2 and pH 7 (Fig. 3.7). Here, for all swelling and shrinking cycles, response times with a small

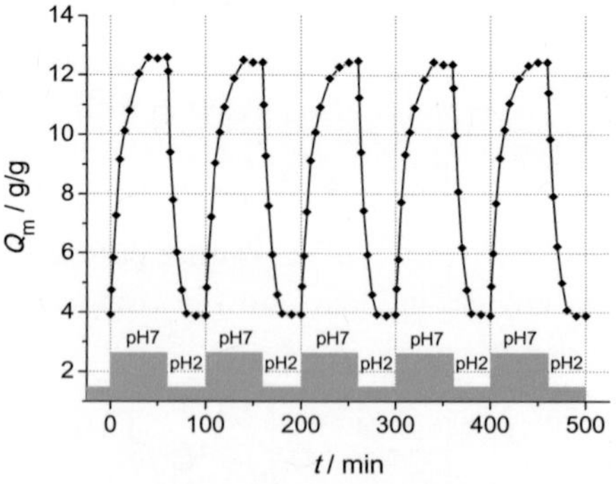

Figure 3.7: Time-dependent mass swelling degree of a PEG6-32 hydrogel sample for cyclic pH changes between pH 2 and pH 7 (sample diameter 1 mm).

standard deviation has been found, i.e. (27 ± 1) min and (15 ± 1) min, respectively. Hence, it can be concluded that the pores do not collapse during the shrinking cycle thus that fast reswelling can occur with excellent reversibility and repeatability.
The use of PEG-6000 as a porogen for the fabrication of porous p(AAm-*co*-AA) hydrogels has been found to lead to a significant response time reduction and to reversible swelling/de-swelling characteristics of the synthesized hydrogels. Thus, among the porogens studied here, this porogen is best-suited for the synthesis of fast-responsive pH sensing materials. Therefore, PEG6000 with a weight ratio of 32 wt% (PEG6-32) is studied in detail in the following.

3.2.2 ATR-FTIR spectroscopy

The formation of an interconnected open-porous network structure requires the complete removal of the porogen after polymerization. To determine if PEG is successfully leached out after washing for five days, freeze-dried hydrogel samples before and after washing have been studied by ATR-FTIR spectroscopy (Bruker Vertex 80v with Golden Gate unit, SPECAC ATR unit, mercury cadmium telluride (MCT) detector, resolution $4\,cm^{-1}$, 100 scans per measurement). Here, the C-O-C asymmetric stretching vibration (ν[C-O-C]) in the range between $1070\,cm^{-1} \ldots 1140\,cm^{-1}$ can be assigned to PEG (cp. Fig. 3.5) and can be used for evaluation [137,138] (Fig. 3.8). From Fig. 3.8 a distinct peak at $1100.4\,cm^{-1}$ can be observed for the porous hydrogel sample before washing (PEG6-32). This clearly indicates the presence of PEG incorporated into the hydrogel. The comparison of the spectra of a non-porous hydrogel (NP) and a porous hydrogel after washing for five days (PEG8-32) reveals the same peaks for both samples and shows that the peak at $1100.4\,cm^{-1}$ completely vanished. Thus, it can be concluded that PEG has been completely removed from the sample after washing.

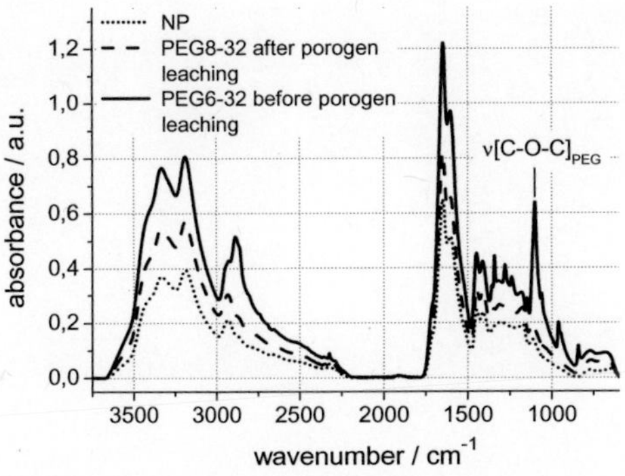

Figure 3.8: ATR-FTIR spectra of conventional non-porous hydrogels as well as porous hydrogels before and after leaching of the porogen by deionized water.

3.2.3 SEM measurements

To directly evaluate the existence of a porous network structure and to analyze and compare the morphology of conventional and porous gel networks, the synthesized cylindrical hydrogel samples were freeze-dried (from the water-swollen state), cut into slices, and studied by SEM (Zeiss Neon 40 EsB). The SEM image of a conventional non-porous gel (Fig. 3.9a) exhibits a smooth dense surface where no pores can be observed. The wrinkles at the surface may result from freeze-drying. In contrast, the SEM image of a PEG6-32 sample at the same magnification clearly shows a porous network structure (Fig. 3.9b). Here, a regular pore distribution with circular pores of different diameter can be observed. The pore diameters were found to vary between 400 nm and 5 µm (see Fig. A.1 in Appendix A.1.2) and the images suggest that pores in the µm range are predominant. The pore walls are very thin and are in the sub-µm range. Furthermore, the inset in Fig. 3.9b reveals that the single pores are connected to each other which strengthens the assumption that an open-porous interconnected network is formed. These results suggest that channels for fast convective solution transport and consequently thin pore walls, determining the critical dimension for cooperative diffusion, have been successfully created.

The comparison of the pore structure of hydrogel samples synthesized in the presence of different PEG concentrations shows a quite similar sponge-like morphology with comparable pore diameters (Fig. 3.10). This is in very good agreement with the free-swelling experiments where no further response time reduction has been observed for increasing PEG concentration. Additionally, all SEM images reveal clean pores where no residues of the porogen can be observed. This confirms the results obtained by ATR-FTIR spectroscopy. However, Figs. 3.10a-c suggest that the polymer fraction is

(a)

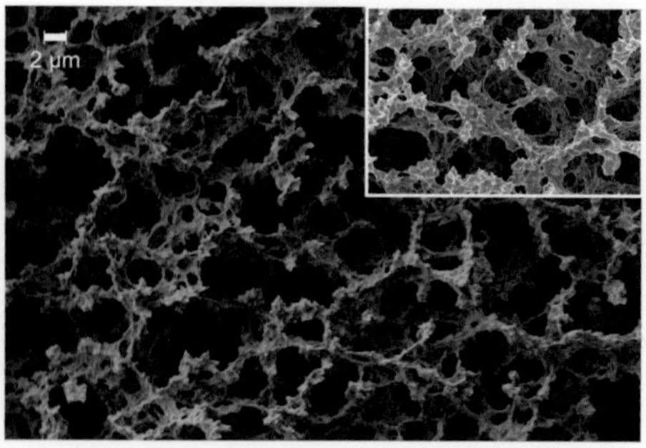

(b)

Figure 3.9: SEM images of freeze-dried p(AAm-*co*-AA) hydrogels (magnification 2×10^3): (a) conventional non-porous hydrogel (NP) and (b) porous PEG6-32 hydrogel. The inset in (b) shows a larger scale image recorded using the in-lens detector of the microscope (magnification 5×10^3).

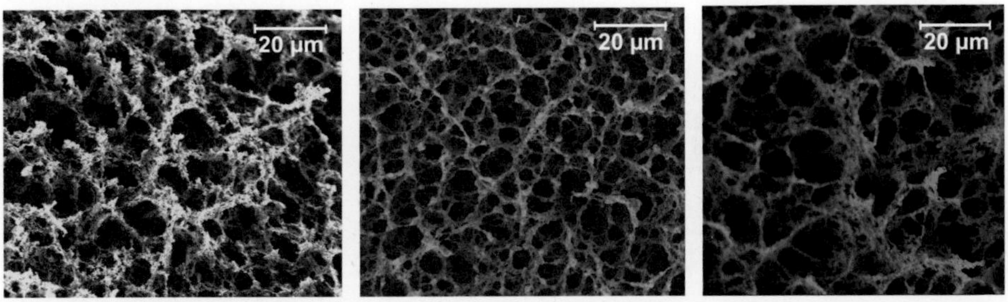

(a) (b) (c)

Figure 3.10: SEM images of freeze-dried p(AAm-*co*-AA) hydrogels (magnification 1×10^3): (a) PEG6-32, (b) PEG6-50, (c) PEG6-80.

decreasing with increasing PEG concentration which reflects in a decreasing thickness of the pore walls. This might reduce the mechanical stability of the gel network which is disadvantageous with respect to the application in piezoresistive microsensors. Therefore, PEG6-32 hydrogels are considered as a good compromise between the existence of an interconnected network structure (cp. free-swelling experiments) and a sufficiently stable polymer scaffold and are thus used for further studies.

3.2.4 Mercury intrusion porosimetry

Since a porous network structure could be proved by SEM imaging, the exact determination of the pore size and pore size distribution of the bulk material is now carried out by mercury intrusion porosimetry (MIP). With this technique, mercury is intruded into the hydrogel sample at specific pressures and the intruded mercury volume, which corresponds to the pore volume, is measured. The obtained data is used to determine the relation between pore size and pore volume by means of the Washburn equation [139]

$$d_{\text{pore}} = -\frac{4 \cdot \gamma \cdot \cos\theta}{p}. \tag{3.3}$$

Here, p is the applied pressure necessary to intrude mercury into a pore of diameter d_{pore}, γ is the surface tension of mercury ($\gamma = 0.485\,\text{Nm}^{-1}$ at 20 °C), and θ is the contact angle between mercury and the hydrogel surface. For the contact angle of mercury on the polymeric p(AAm-*co*-AA) surface, a value of $\theta = 140\,°$ was used for calculations [140, 141]. However, note that Eq. (3.3) was originally derived for cylindrical pores and is thus only an estimate for the determination of the pore diameter if the pore geometry differs from ideal cylindrical shape. For MIP measurements (Porosimeter 2000, max. pressure 2000 bar), cylindrical non-porous

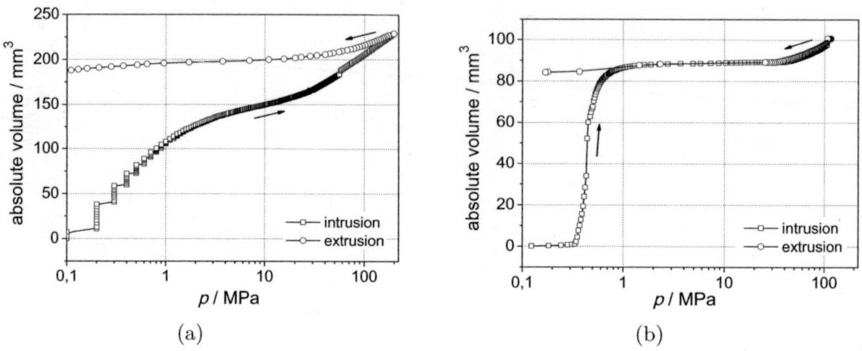

Figure 3.11: Intrusion-extrusion isotherms of (a) non-porous (NP) and (b) porous (PEG6-32) hydrogels measured by mercury intrusion porosimetry.

(NP) and porous (PEG6-32) hydrogels have been freeze-dried from the water-swollen state, cut into 10 mm pieces and inserted into the penetrometer.

When the pressure is gradually increased, the pores are filled with mercury which reflects in an increase in the absolute mercury volume. For non-porous hydrogel samples, the intrusion curve does not show a sharp transition but a smoothly increasing behavior with two transition regions (Fig. 3.11a). This suggests that at pressures below 1 MPa, the voids between the different sample pieces and the pores formed during freeze-drying are filled. At pressures above 10 MPa, smaller pores which are either intrinsically existent in the gel or which are formed by air bubbles during the exothermic gel formation are invaded. The extrusion curve shows a distinct hysteresis behavior, indicating that mercury is not extruded when the pressure is lowered. In contrast, the intrusion curve for a PEG6-32 sample shows an almost abrupt increase at a pressure of 0.33 MPa and reaches a plateau at a pressure of about 1.3 MPa (Fig. 3.11b). This suggests a completely different hydrogel morphology compared to conventional gels. Here, it can be assumed that almost all pores inside the sample are filled within a narrow pressure range which leads to the saturation of the intrusion curve for pressures above 1.3 MPa. The upwarding characteristics of the intrusion curve for pressures above 38 MPa can be attributed to compression of the sample due to the high pressure and does not originate from small pores filled by mercury [142]. The extrusion curve for a decreasing pressure first follows the path of the intrusion curve but a permanent hysteresis can be observed as well. This phenomenon is well known for MIP and can be explained by the existence of pores with a large pore volume-to-pore throat ratio ("bottle neck" pores) or rupture of the mercury in pore constrictions or junctions during extrusion [143, 144].

From Eq. (3.3), the pore size distribution can be calculated and it can be seen that

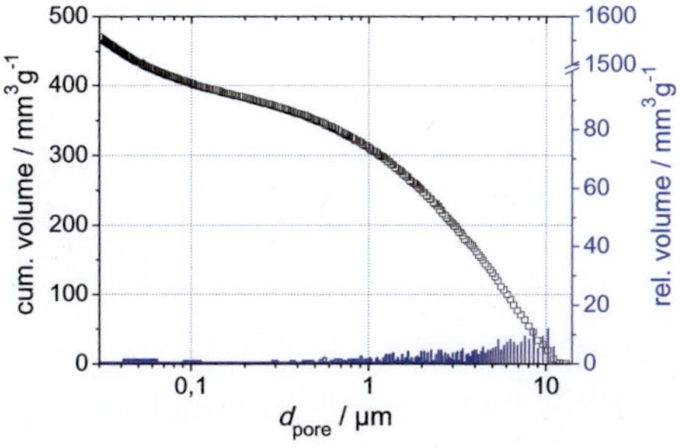

(a)

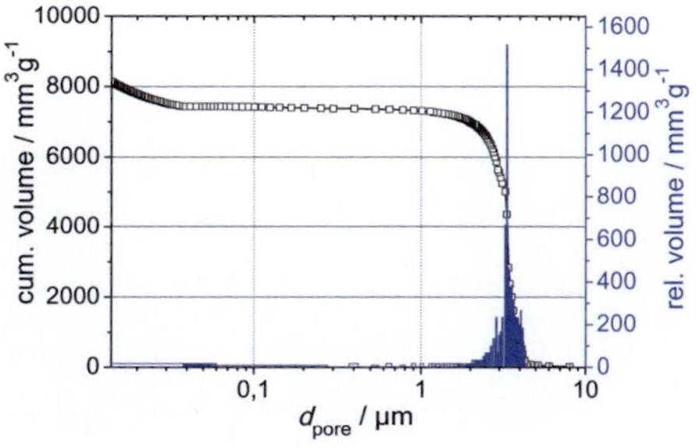

(b)

Figure 3.12: Cumulative intrusion volume and pore size distribution of (a) non-porous (NP) and (b) porous (PEG6-32) hydrogels measured by mercury intrusion porosimetry.[2]

the applied pressure and the pore diameter are inversely proportional, i. e. at small pressures, larger pores are filled and with increasing pressure smaller pores are invaded which reflects in the curves for the cumulative volume (Fig. 3.12). For conventional non-porous samples, a wide pore size distribution can be observed (Fig. 3.12a). The small pores below 0.1 µm might be attributed to the intrinsic interstices inside the gel. The large pores between 1 µm and 10 µm can be assigned to the altered sample surface by freeze-drying or voids between different samples. For PEG6-32 hydrogels, a narrow pore size distribution between 2 µm and 5 µm can be observed (Fig. 3.12b). Here, an average pore diameter of 3.3 µm has been found. This value is in very good agreement with the results obtained by SEM and confirms the successful formation of a macroporous network structure.

3.3 Sensor integration

Since a macroporous network structure of PEG6-32 hydrogels has been proved as well as a significant response time reduction for free-swelling bulk gels, sub-100 µm sensing layers of PEG6-32 materials are fabricated for integration into piezoresistive microsensors. The fabrication process, the layer integration, and the sensor set-up are described in the following.

3.3.1 Sensor layer fabrication

To facilitate the fabrication of reproducible sensing layers for integration into microsensors, macroporous hydrogels have been attached to standard glass substrates

[2]The cumulative and the relative volume are related to the net weight of the hydrogel samples used for MIP and depend on the chosen pressure steps. Thus, the absolute values at the ordinates have no actual meaning and should not be directly compared.

by silane chemistry. Surface modification of the substrates has been done by the following steps:

1. Cleaning of the glass substrates in deionized water and ethanol, respectively (ultrasonic bath, 30 min);
2. RCA clean [145], flushing with deionized water, thermal annealing (1 h, 120 °C);
3. Exposure of the clean substrates to a 0.4 vol% silane solution (640 µl 3-methacryloxypropyltrimethoxysilane (MPS) and 160 ml ethanol) for 12 h under stirring;
4. Thermal annealing (10 min, 150 °C).

In this way, a reactive layer which is bound to the glass substrate on one side and has reactive double bonds on the other side is established (step #2 in Fig. 3.13). The bonding process of the hydrogel to the substrate occurs during the polymerization reaction (step #4 in Fig. 3.13).

To fabricate homogeneous sensing layers with defined thickness, the modified glass substrates have been attached to unmodified glass slides in a defined distance using spacers (height 200 µm). The spacers have been arranged such that an open channel between both glass slides exists. The channel has been flushed with nitrogen. The PEG6-32 pre-gel solution (see Section 3.1.3) is brought to the channel opening such that the channel is homogeneously filled using capillary forces (step #3 in Fig. 3.13).

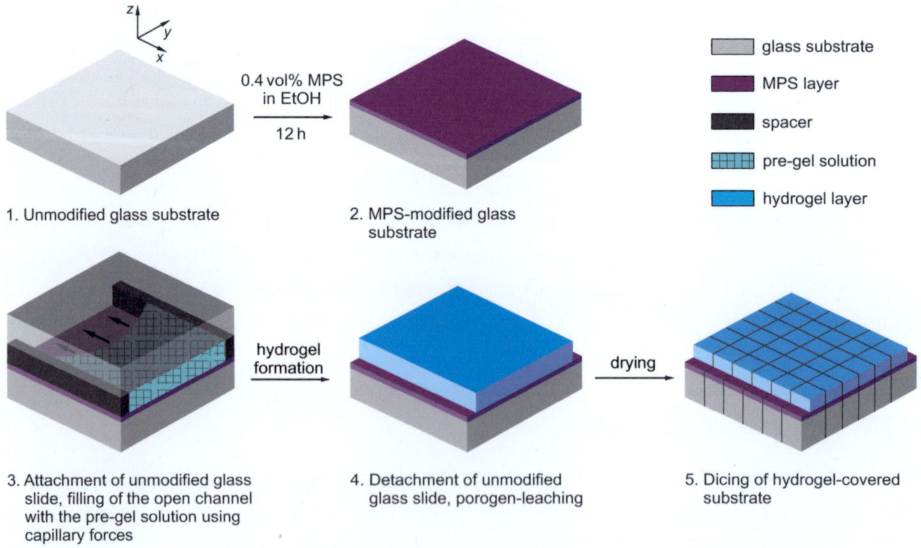

Figure 3.13: Process for immobilization of defined macroporous p(AAm-*co*-AA) sensing layers on a glass substrate (not drawn to scale).

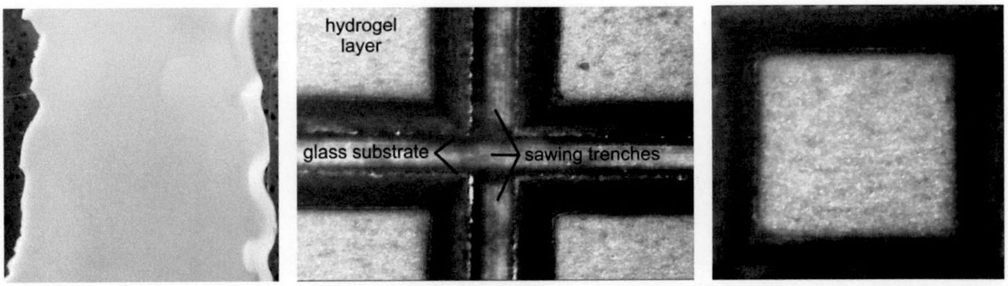

(a) (b) (c)

Figure 3.14: Macroporous PEG6-32 hydrogel layers attached to a glass substrate: (a) photograph of the closed layer after fabrication, (b) optical microscope image of four single dies after sawing, (c) $(1 \times 1)\,\mathrm{mm}^2$ single die.

To allow complete polymerization, the unmodified glass slides and the spacers have been removed after 24 h by gently flushing with deionized water (step #4 in Fig. 3.13). After this process, a homogeneous hydrogel layer with a nontransparent white color could be observed. This indicates that phase separation and, thus, the formation of a porous network structure has been successfully established also for thin sensing layers (Fig. 3.14a). It has been found that over 95% of the hydrogel layers are perfectly attached to the substrate after the fabrication process and that just the edges of the layers, which have been in direct contact with the surrounding air, are slightly corrugated. It is assumed that the presence of oxygen results in an incomplete polymerization and bonding process.

Washing of the sensing layers in water for five days to remove PEG, revealed a stable attachment of the gel layers to the substrate. Even after 14 days in water, more than 90 % of the gel layer was still bound to the substrate. However, alternative leaching of PEG with PBS buffer solution (pH 7.4, $I = 0.15\,\mathrm{mol/l}$) showed a progressive peel-off of the hydrogel layer beginning from the edges after three days and complete detachment after six days. In both cases, due to the attachment of the hydrogel layer to the glass substrate, isotropic swelling is restrained and a bidirectional in-plane swelling stress σ_{xx}, σ_{yy}, acting on the hydrogel layer parallel to the glass surface, exists. The bidirectional swelling stress (normal stress) can be written as [146]

$$\sigma_{xx} = \sigma_{yy} = kT\nu_{\mathrm{c}}\left(\varphi_{\mathrm{p,eq}} - \frac{1}{\varphi_{\mathrm{p,eq}}}\right) \tag{3.4}$$

and, thus, depends on the cross-linking density ν_{c} and the polymer volume fraction $\varphi_{\mathrm{p,eq}}$ at equilibrium. The swelling of the gel layer in PBS is more pronounced compared to water because the carboxyl groups of AA are in the deprotonated state leading to a higher gel hydrophilicity and, thus, to a decrease of $\varphi_{\mathrm{p,eq}}$. Hence, the absolute value of the in-plane swelling stress increases and leads to progressive detachment of the gel layer. Furthermore, due to the high porosity of the hydrogel, it can be assumed that the total polymer area, which can bind to the MPS adhesion layer, is reduced compared to non-porous hydrogels because less binding sites are available. This decreases the critical stress which can be applied to the gel layer before detachment is initiated. Therefore, the reliable binding of porous hydrogels to rigid substrates to sustain high in-plane swelling pressures should be addressed in future research.

After porogen leaching, the glass substrates with attached hydrogel layers have been dried at room temperature for about 48 h and diced to single chips of $(1 \times 1)\,\text{mm}^2$ using a wafer saw (step #5 in Fig. 3.13, Figs. 3.14b,c). The thickness of the dried hydrogel layers was about 70 µm.

3.3.2 Sensor set-up

According to the general principle of hydrogel-based sensors (see Fig. 1.2 in Section 1.3.1), the sensor is build up on a pH-sensitive macroporous p(AAm-*co*-AA) hydrogel layer attached to a glass substrate and a square piezoresistive pressure sensor chip (Fig. 3.15). The glass die with attached hydrogel layer is mounted to the center of a modified TO8 package containing fluidic channels for supplying the measuring solution. The pressure sensor chip is then attached to a TO8 package using an epoxy-based adhesive such that the hydrogel layer is sandwiched between

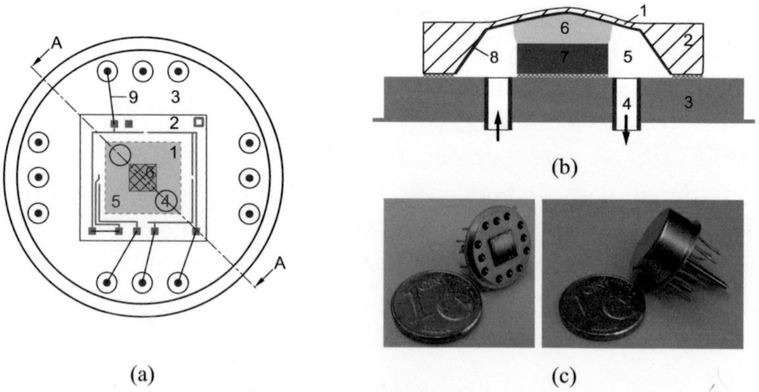

Figure 3.15: Sensor setup of a piezoresistive pH microsensor containing a macroporous p(AAm-*co*-AA) sensing layer: (a) top view (with half-transparent flexure plate), (b) A-A cross sectional view, and (c) photographs of the sensor with and without protection cap. 1 silicon flexure plate (thickness 20 µm), 2 silicon pressure sensor chip ($5.2 \times 5.2 \times 0.4\,\text{mm}^3$), 3 modified TO8 package, 4 fluidic channels, 5 pH measuring solution, 6 hydrogel layer, 7 glass substrate, 8 Si_3N_4 passivation layer, 9 bonding wire.

the glass substrate and the flexure plate (Fig. 3.15b). The pressure sensor chip is fabricated by conventional micromachining techniques, i. e. wet etching from the chip backside using a silicon nitride mask as etch resist and backside passivation by means of a 200 nm silicon nitride film deposited by plasma enhanced chemical vapor deposition (PECVD). The latter enables good protection of the transducer in harsh environments as well as prevention of ion diffusion [147, 148]. The measuring solution is continuously supplied using a syringe pump (GeSIM GmbH, Grosserkmannsdorf). All measurements have been carried out in a temperature chamber (Heraeus-Vötsch) at a constant controlled temperature of 20 °C.

3.4 Sensor characterization

Sensors with integrated macroporous hydrogel layers are characterized with respect to their pH response in terms of sensitivity, time-dependent sensor response, and repeatability and are directly compared to sensors exhibiting conventional non-porous sensing layers.

3.4.1 Sensor conditioning

In order to bring the sensor into a defined reference state, sensor conditioning, i. e. the application of pH 3 buffer solution to the sensor, containing a dry hydrogel layer, has been performed (Fig. 3.16). The supply of the measuring solution to the dry gel leads to swelling of the hydrogel layer and, consequently, to a fast increase in the sensor output signal indicating a proper sensor operation. However, from Fig. 3.16 a two-step process can be observed, i. e. after about 3 min, before the output signal has reached its steady state, a sudden decrease of V_{out} can be observed until the

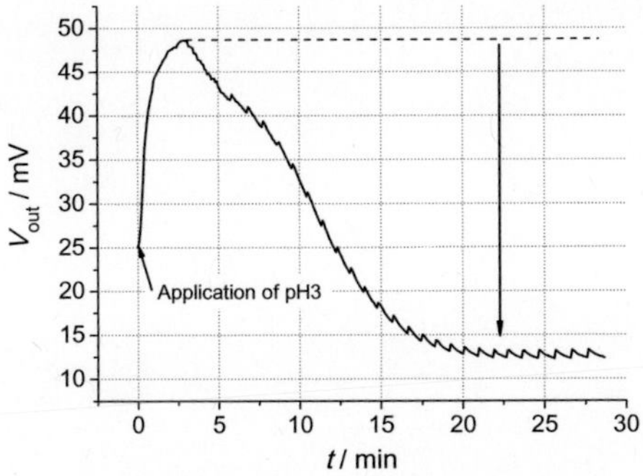

Figure 3.16: Time-dependent electrical sensor output signal of a sensor with integrated PEG6-32 hydrogel layer for the conditioning process in pH 3 buffer solution ($I = 0.2$ mol/l).

sensor signal equilibrates after about 20 min. The corresponding output voltage in the steady state appears to be below the sensor offset voltage of about 25 mV. This implies that the deflection of the flexure plate is smaller compared to the deflection directly after sensor assembly. The solvent uptake of the hydrogel decreases the polymer volume fraction and, thus, the elastic modulus of the sensing layer. Here, the porous network structure is assumed to additionally weaken the mechanical stability of the gel matrix such that the restoring force of the flexure plate has a non-negligible influence on the sensor behavior. Thus, it can be assumed that the macroporous gel layer is compressed by the restoring force of the flexure plate and evades horizontally into open spaces. This may lead to the compression of the pore structure and to a decrease of the effective pore radius. This assumption is supported by the fact that such a behavior has not been observed for free-swelling porous gels nor for sensors containing non-porous hydrogels. This phenomenon has not yet been verified by other methods but is an undesired sensor behavior whose solution should be concerned in future research.

3.4.2 Transfer characteristics

Starting from the reference state after sensor conditioning, pH has been increased step-wise up to pH 7 to determine the sensor transfer characteristics (Fig. 3.17a). Here, an increasing output signal with sigmoid shape for increasing pH can be observed. For pH values below pH 4, the carboxyl groups of AA exist in their undissociated, protonated form such that the gel hydrophilicity is low. This leads to a small swelling degree of the sensing layer and consequently to a small output voltage. The steep slope of the characteristic curve between pH 4 and pH 5 can be attributed to the

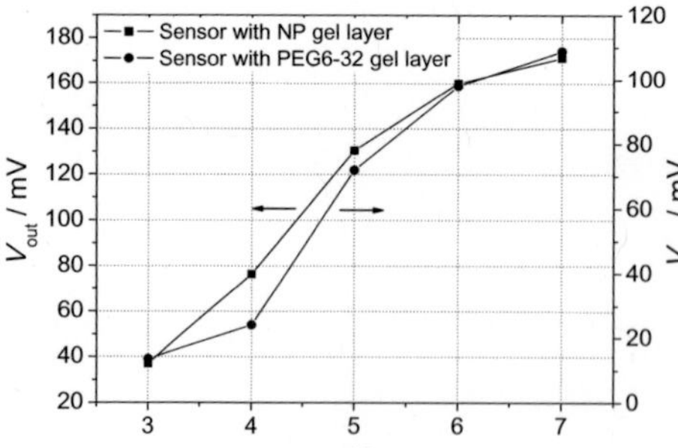

(a)

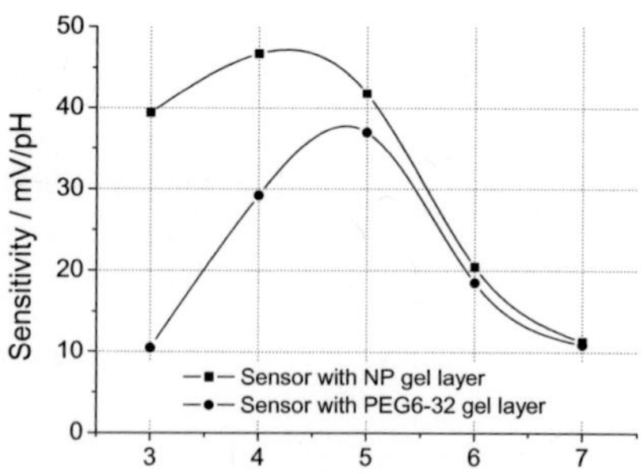

(b)

Figure 3.17: (a) Transfer characteristics and (b) sensitivity as a function of pH for sensors with macroporous (PEG6-32) and non-porous (NP) hydrogel layers, respectively ($I = 0.2\,\text{mol/l}$).

dissociation of the carboxyl groups around the acid dissociation constant of AA ($pK_a \approx 4.7$). Thus, the charged carboxylate groups – fixed on the polymer chains – repel each other and the gel hydrophilicity is increased. This leads to an enhanced solvent uptake of the sensing layer and, thus, to an increase of the sensor output signal.

A direct comparison between sensors containing macroporous and sensors containing conventional non-porous hydrogel layers reveal qualitatively similar transfer characteristics. This can be expected since the introduction of pores only alters the hydrogel morphology but the pH response only depends on the chemical nature of the sensitive groups attached to the polymer backbone. These groups are the same for both types of gels. However, the absolute output voltage for a certain pH value as well as the sensor sensitivity is smaller for sensors containing macroporous hydrogels compared to sensors with non-porous gel layers (Figs. 3.17a,b). Here, reduced output signal values and a reduced sensitivity of up to a factor of four were found. The reason for the smaller signal amplitudes might be a reduced hydrogel stability such that the restoring force of the silicon flexure plate suppresses the swelling pressure of the gel as observed for sensor conditioning. Furthermore, hydrogels synthesized in the presence of porogens exhibit a reduced polymer volume fraction because the created channels can just be filled with solvent molecules but do not contain ionizable functional groups responsible for hydrogel swelling. This reduced the overall sensitivity of these gels. An improved sensor sensitivity might be achieved by increasing the number of functional groups, i. e. the ratio of AA in the pre-gel solution or by producing hydrogels with enhanced elastic modulus.

3.4.3 Time-dependent sensor response

To study the influence of the macroporous gel morphology on the sensor response time, the time-dependent sensor output signal has been measured for different pH steps (Fig. 3.18). For a pH step from pH 3 to pH 6, an increasing sensor signal can be observed for both types of sensors (Fig. 3.18a). Here, the time to reach equilibrium and the sensor response time have been found to be reduced for sensors containing macroporous sensing layers. In particular, a response time t_{90} of about 17 min has been found for sensors with PEG6-32 gel layer. This corresponds to a response time reduction of about 60 % compared to sensors with non-porous hydrogel layers. For deswelling, the same qualitative behavior has been found with a response time of about 11 min which is a reduction of about 55 % (Fig. 3.18b). This indicates that the sensor response is fully reversible and a response time reduction could be achieved for swelling and deswelling. The smallest response time observed was for a pH step from pH 7 to pH 8 with a response time t_{90} of about 6.5 min (Fig. 3.18a). This is especially

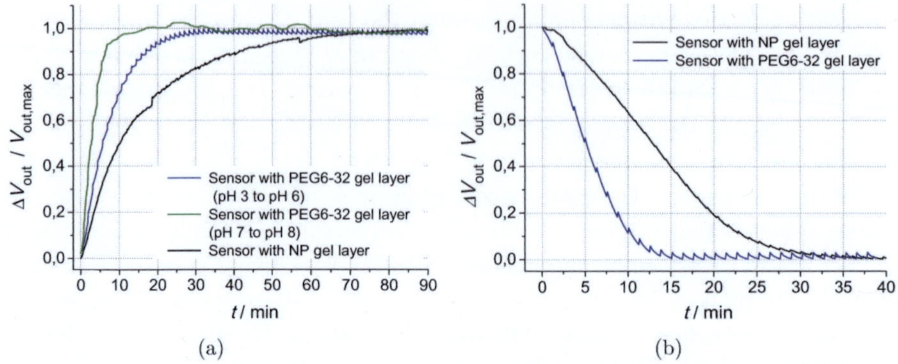

Figure 3.18: Normalized time-dependent sensor output voltage for pH steps between (a) pH 3 and pH 6 as well as pH 7 and pH 8 and (b) pH 6 and pH 3 ($I = 0.2\,\text{mol/l}$).

advantageous for sensors used for monitoring of physiological fluids with pH values around 7.4. The same improved sensor behavior has been observed for different pH step heights and different pH ranges (Fig. 3.19a). A pulse-wise change of the pH value between pH 7 and pH 5 revealed a stable sensor output signal and excellent repeatability where no long-term drift of the signal amplitude could be observed (Fig. 3.19b). This is in good agreement with the results obtained for free-swelling gels exited by pulse-wise pH changes (see Section 3.2.1c).

Despite the improved sensor response for sensors containing macroporous gels, the response time reduction was expected to be more pronounced based on the results obtained from free-swelling experiments (cp. Section 3.2.1). This corresponds to the

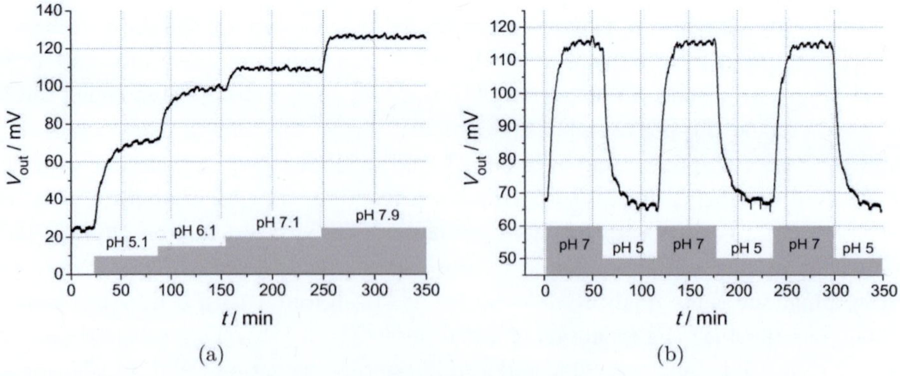

Figure 3.19: Time-dependent sensor output signal: (a) for incremental pH steps between pH 4 and pH 7.9, (b) for pulse-wise pH excitation between pH 7 and pH 5 ($I = 0.2\,\text{mol/l}$).

hypothesis that the pores are compressed by the flexure plate during the conditioning process. It may be also possible that the porous network structure has been partly destroyed leading to a denser network with hindered solution transport. This should by verified in detail in further studies.

3.5 Conclusions and summary

In this chapter, for the very first time, the application of macroporous pH-sensitive p(AAm-*co*-AA) hydrogels as smart sensing layers into piezoresistive microsensors has been described. For the fabrication of a porous network structure with interconnected pores in the µm-range, the porogen-leaching method has been used. In particular, four different porogens, namely NaCl, $CaCO_3$, 1,4-dioxane, and PEG have been applied to synthesize porous hydrogels with improved swelling/shrinking rates. The time-dependent pH response of these gels was determined by free-swelling experiments and has been directly compared to evaluate the porogen material which works best. It has been found that hydrogels synthesized in the presence of PEG with a molecular weight of $6000\,g\,mol^{-1}$ (PEG6000) showed the fastest swelling response. For the use of PEG with lower molecular weight (PEG1000), no enhancement of the swelling rate could be observed, whereas the use of PEG with higher molecular weight (PEG8000) did not show further improvements of the hydrogel response compared to PEG6000. Furthermore, it has been found that a PEG ratio of 32 wt% is sufficient for the creation of an open porous network structure.

ATR-FTIR spectroscopic characterization revealed that PEG could be completely removed from the polymer network by washing in water for 5 days. This was confirmed by SEM imaging where the formation of an open porous network structure with thin pore walls in the sub-µm range was clearly shown. The pore diameters estimated from the SEM images were found to be between 400 nm and 5 µm. Mercury intrusion porosimetry measurements allowed the investigation of the exact morphological features, i. e. pore size and pore size distribution, and revealed a narrow pore size distribution with an average pore diameter of about 3.3 µm which perfectly confirmed the results obtained by SEM.

For the integration into piezoresistive microsensors, these hydrogels have been successfully attached to silane-modified glass substrates and single dies have been produced by sawing. Porogen-leaching in water showed good stability of the attached hydrogel layers. However, the application of PBS buffer solution at pH 7.4 revealed poor stability. It was assumed that the chemically promoted adhesion of the layer is too weak to withstand the high bidirectional swelling stress which might be additionally supported by the reduced number of bonding sites due to the porous network structure. This is a serious drawback regarding the long-term stability of these sensing layers and should be addressed in future research.

The integration of these hydrogel structures into piezoresistive microsensors revealed sensors with a distinct pH response. Sensors based on macroporous hydrogels showed a slightly reduced absolute sensor signal and smaller sensitivity compared to sensors containing non-porous hydrogels. This could be explained by the reduced polymer volume fraction and the smaller number of pH-sensitive functional groups due to the porous structure. Furthermore, this is assumed to be additionally supported by the restoring force of the flexure plate. This force is assumed to have a non-negligible influence on the sensor response due to a reduced elastic modulus of the porous network. The time-dependent sensor response revealed the successful achievement of a response time reduction for swelling and deswelling. Here, for a pH change between pH 3 and pH 6 and vice versa, a response time reduction of 60 % and 55 % has been found, respectively. The smallest response time observed was about 6.5 min for a pH step between pH 7 and pH 8. For all sensors, a stable output signal with good repeatability and reproducibility could be achieved with the technology used.

4 Sensors based on clay nanocomposite hydrogels [1]

In this chapter, novel photopolymerizable organic/inorganic nanocomposite hydrogels consisting of PNIPAAm and Laponite XLS clay nanoparticles are studied with respect to their application as smart sensing layers in piezoresistive microsensors. At first, the structural and mechanical properties of these sensing layers are investigated by different analysis techniques and are compared to the properties of conventional PNIPAAm gels. Subsequently, both types of gels are integrated into the microsensor devices and the sensor properties are directly compared and discussed.

4.1 PNIPAAm hydrogel networks

4.1.1 Sensoric properties

PNIPAAm hydrogels, cross-linked with organic crosslinkers like BIS (BIS-gels), are among the most studied non-ionic polymer systems and are specifically known for their thermosensitivity. Similar to un-cross-linked PNIPAAm which shows a lower critical solution temperature (LCST), the hydrogel undergoes a phase transition at a temperature of about 32 °C below which the gel is in its swollen state and above which the network structure is collapsed [149]. This discontinuous phase transition behavior is associated with the phase separation of single chain polymers in aqueous solutions and can be attributed to a sensitive equilibrium of hydrophilic and hydrophobic groups of the polymer [150, 151] (Fig. 4.1). The location of the LCST is determined by the total hydrophilicity of the polymer. Thus, an increase of the hydrophilic groups leads to a decrease of the phase transition temperature. The thermo sensitive properties can be advantageously employed in copolymer sensing layers consisting of PNIPAAm and a pH-sensitive component to enhance the reliability and long-term stability of pH sensors by systematic temperature stimulation during the regeneration cycles [103].

Despite the water–polymer interaction, more general, the polymer–water–analyte interaction determines the swelling state of the gel. In particular, by addition of an analyte to the hydrated gel, the water–polymer hydrogen bonding is perturbed leading to a change in the phase transition temperature and consequently to change

[1] This chapter is the result of a close cooperation with the Professorchip for Physical Chemistry and Physical Chemistry of Polymers, TU Dresden. Especially hydrogel synthesis was carried out in cooperation with Dr. B. Ferse and A. Grosse.

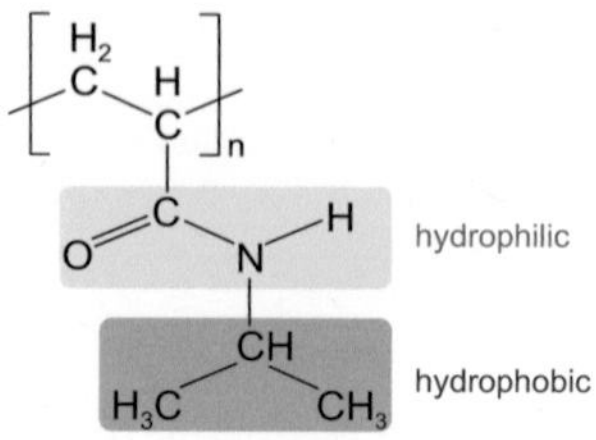

(a)

Figure 4.1: Chemical structure of PNIPAAm.

in the swelling state. This is the case for organic co-solvents like ethanol, methanol, 1-propanol etc., or different salt ions, such that PNIPAAm gels can be well applied as sensing layers in chemical sensors for detection of these analytes [116].

4.1.2 Conventional gels

Conventional PNIPAAm gels, i. e. chemically cross-linked hydrogels like poly(NIPAAm-*co*-BIS) show very poor mechanical properties. In particular, the existence of randomly oriented covalently cross-linked polymer chains leads to stiffening of the polymer segments and to an irregular network structure which leads to either a very soft (low cross-linking density) or a very brittle (high cross-linking density) network. This is a serious drawback regarding the necessity of reliable, reproducible and long-term stable force transduction to the flexure plate of the pressure sensor chip. With increasing crosslinking density and the corresponding stiffening of the network, the swelling degree decreases. Thus, for sensor applications, complementarity between a stable network and a high swelling degree, i. e. high sensitivity, does exist. Therefore, only a very limited range of crosslinker concentrations can be used. Moreover, the incorporation of the crosslinker BIS leads to altered sensor properties with increasing BIS concentration, e. g. to a shift of the phase transition temperature.

For gels which are suitable for sensor applications, regarding the crosslinker concentration and the layer thickness, another phenomenon, known as shrinkage barrier effect, occurs. This effect can be observed during the shrinking cycle of the hydrogel and can be attributed to the formation of a skin-like collapsed layer at the surface of the hydrogel [152]. This dense layer inhibits the release of the solution inside the gel and causes very long sensor response times for deswelling. This effect can be observed for deswelling due to temperature increase and for addition of organic co-solvent concentrations below 50 vol% to the water-swollen gel.

4.1.3 Strategies for preparation of mechanically stable hydrogels

In general, physical networks are suitable for the preparation of mechanically stable hydrogels. An overview of different preparation methods can be found in [153] and can be briefly summarized as:

- *Double-network gels*: Two networks interpenetrate each other such that an insoluble collective network develops [154].
- *Slide-ring gels*: These gels are cross-linked via figure-of-eight crosslinkers, where one strategy is the threading of polyethylene glycol chains through α-cyclodextrin-rings and subsequent crosslinking as shown by OKUMURA *et al.* [155]. Under deformation, the polymer chains maintain their flexibility by sliding through the rings ("topological gels").
- *Nanocomposite gels based on layered silicates*: This method was first proposed by HARAGUCHI *et al.* [156] and uses layered silicates as multifunctional crosslinks between polymer chains. During the polymerization reaction, the monomers are adsorbed to the silicate surface and the polymer chains develop between the surfaces of the layered silicate particles resulting in a strong physical interaction between both. These interactions lead to superior mechanical properties.

4.2 Clay nanocomposite hydrogels

Not only due to their extraordinary mechanical properties but also due to the simple preparation procedure, especially nanocomposite hydrogels are suitable candidates for sensor applications. In contrast to the hydrogels synthesized by HARAGUCHI *et al.*, where the free radical polymerization is initiated with potassium peroxodisulfate (KPS) and TEMED, photochemical initiation is used for the hydrogels studied in this work. Hydrogels prepared in this way have been recently reported by FERSE *et al.* [157,158] for the first time. The photopolymerization technique at a wavelength of approximately 360 nm is appropriate for obtaining well-defined hydrogel structures in the micrometer range and can be easily integrated within MEMS processes.

4.2.1 Layered silicates

Layered silicates, known from nature and found in clay minerals such as mica or talc, can be classified in cation-poor and cation-rich silicates. In both, the silicate layer, consisting of corner-connected SiO_4 tetrahedrons (Fig. 4.2a), is corner-connected with a octahedral cationic layer (Fig. 4.2b). The latter mainly incorporates Mg^{2+}, Ca^{2+}, Al^{3+}, and Fe^{3+} cations and these layers are formed by edge-connected single octahedrons. The connection between the tetrahedral and octahedral layers can

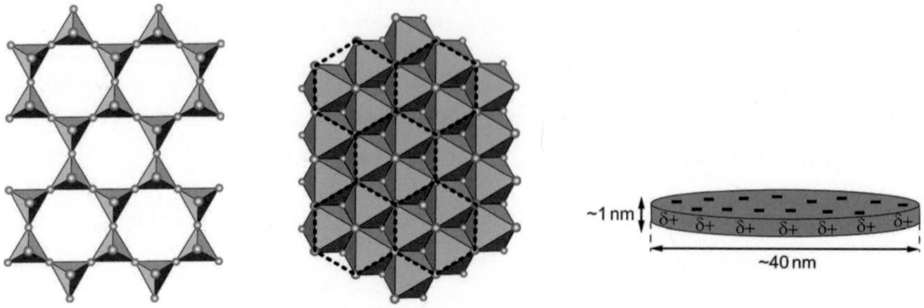

(a) (b) (c)

Figure 4.2: (a) Silicate layer consisting of edge-connected SiO_4 tetrahedrons, (b) edge-connected octahedral cationic layer [159], (c) size and shape of a Laponite nanoparticle. For Laponite, the octahedral layer is sandwiched between two tetrahedral layers. The atoms shown in (a) and (b) are oxygen atoms; silicon atoms are located inside the tetrahedrons in (a) and magnesium and lithium atoms are located inside the octahedrons in (b). The negative charge at the surface is compensated by Na^+ ions (not shown) [153].

develop in two ways: (i) an octahedral layer is connected with only one side to a tetrahedral layer (cation-rich layered silicates), (ii) an octahedral layer is connected to two tetrahedral layers forming a sandwich-like structure (cation-poor layered silicates). Clay minerals like hectorite or montmorillonite are cation-poor silicates and belong to the 2:1 phyllosilicates. The surface of these layered particles is negatively charged whereas the rim is positively charged so that water uptake between the layers is possible [159] and the lamellar crystalline structure is cleaved into disc-like particle upon dispersion in water.
In this work, the inorganic clay Laponite ($(Mg_{5.34}Li_{0.66}Si_8O_{20}(OH)_4)Na_{0.66}$), which is a synthetic hectorite, was used as crosslinker. The Laponite particles have a disc-like shape with a diameter of about 40 nm and a height of about 1 nm (Fig. 4.2c). Two types of Laponite do exist:

- *Laponite XLG*: Dispersions in water lead to the formation of a "house-of-cards" structure of the clay particles when a critical concentration is exceeded ($\approx 2\,$wt%). The positively charged rim and the negatively charged particle surfaces attract each other by electrostatic forces and, thus, a gelation of the aqueous solution takes place [160]. This is disadvantageous with regard to a desired uniform dispersion of clay particles for hydrogel synthesis for higher crosslinker concentrations, so that the use of this Laponite type is less advantageous.

- *Laponite XLS*: This type is similar to Laponite XLG but the positively charged rim of the particles is modified with pyrophosphate ions ($P_2O_7^{4-}$). When dispersed in water, the equal charge of the rims and the silicate surfaces lead to electrostatic repulsion of the particles and, thus, to a uniform particle distribution. Hence, even for high clay concentrations, no gelation occurs and hydrogels with a wide range of crosslinker concentrations can be prepared.

4.2.2 Sensor layer fabrication

For hydrogel synthesis, the NIPAAm monomers, Laponite XLS nanoparticles, and photoinitiator were mixed in water. To study the influence of crosslinker concentration, different pregel solutions with varying clay content have been prepared (Fig. 4.3). To allow direct conclusions of the influence of clay nanoparticles on the sensor behavior, also conventional BIS-gels using the same monomers and photoinitiator have been synthesized. A detailed description of the materials used and the synthesis procedure can be found in [153] and is summarized in Appendix A.2 together with the sample nomenclature.

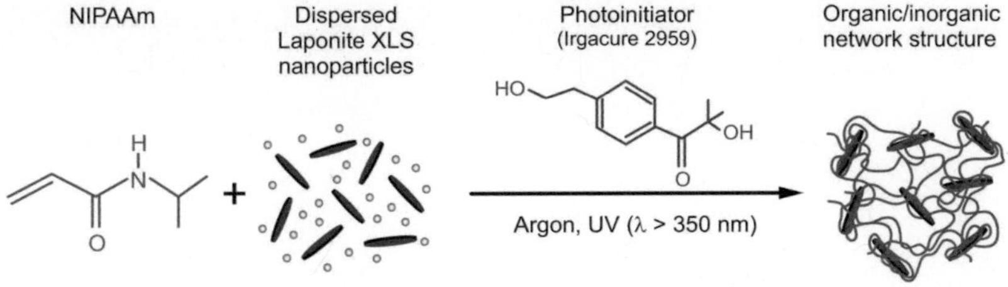

Figure 4.3: Synthesis of hybrid organic/inorganic hydrogels consisting of PNIPAAm and clay Laponite XLS by UV-photopolymerization.

The sensing layers were fabricated as follows:

- The transparent aqueous pregel solutions of BIS- and NC-gels were poured into rectangular molds ($30 \times 35 \times h_\mathrm{m}\,\mathrm{mm}^3$) and covered with a glass plate to enable an exactly defined hydrogel thickness and an even surface after polymerization[2].
- Photopolymerization through UV irradiation under argon atmosphere (mercury UV lamp "LOT Oriel", 900 W, $\approx 5\,\mathrm{mW/cm^2}$, $\lambda > 360\,\mathrm{nm}$, $t_\mathrm{exp} = 120\,\mathrm{s}$).
- Drying of the hydrogel layer at room temperature (layer thickness of the dry gel $d_\mathrm{dry} = (75 \pm 5)\,\mu\mathrm{m}$)
- Cutting the dry hydrogel layer into $(0.9 \times 0.9)\,\mathrm{mm}^2$ pieces using a scalpel.

[2]The height of the molds h_m was varied for different Laponite concentrations such that all layers had a thickness of 75 µm to allow comparability.

4.2.3 Structural characterization

a. Energy-dispersive X-ray analysis

After fabrication of the sensing layers, the incorporation of inorganic clay particles with different concentrations into the sub-100 µm films has been studied by energy dispersive X-ray (EDX) analysis. The comparison of the obtained spectra for conventional BIS-PNIPAAm (Fig. 4.4a) and the nanocomposite gels (Fig. 4.4b) reveal the successful formation of a hybrid organic/inorganic network structure by photopolymerization. The C-, N-, and O-peaks below 60 eV represent the PNIPAAm network whereas the Na-, Mg-, Si-, and P-peaks above 90 eV represent the inorganic clay material. The increase in peak height of the peaks assigned to the clay particles indicate an increase in crosslinking density with increasing clay concentration. The analysis of the peak area for different clay concentrations suggests that the crosslinking density shows a nonlinear dependence on clay concentration and the concentration is significantly higher for NC20-gels. However, a quantitative analysis by EDX is challenging because EDX is a surface-sensitive method. Thus, the results suggest a higher clay content at the layer surface for highly cross-linked gels near the dispersion limit.

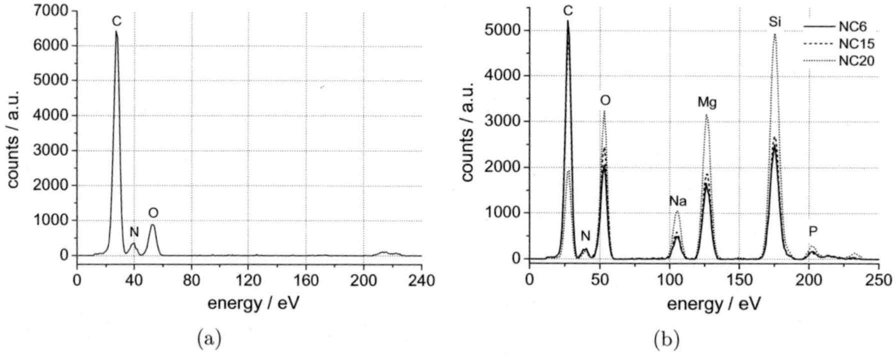

Figure 4.4: EDX spectra of (a) conventional poly(NIPAAM-*co*-BIS) hydrogel layer, (b) PNIPAAm clay nanocomposite hydrogel with different Laponite XLS clay concentrations.

b. Atomic force microscopic imaging

To study the distribution of clay nanoparticles, AFM measurements have been performed on the air-dried sensing layers in tapping mode (Fig. 4.5). During capturing of the height image, a phase image can be recorded simultaneously. The latter provides useful information on the distribution of different materials inside a sample because the phase shift between excitation signal and cantilever oscillation depends on the

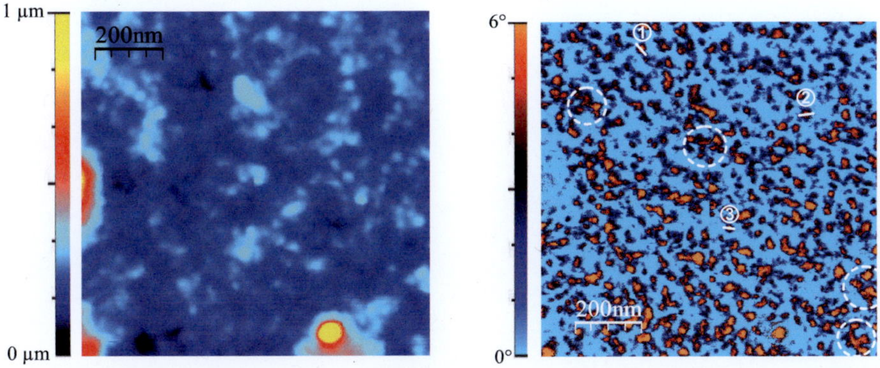

(a) (b)

Figure 4.5: AFM images of an air-dried NC15 hydrogel layer: (a) height image, (b) phase image; tapping mode, image size: $(1 \times 1)\,\mu\text{m}^2$.

elastic/viscoelastic properties, the hardness, and the adhesion properties of the respective surface as [161]

$$\Delta\varphi = \varepsilon \langle r \rangle \frac{E_{\text{eff}} \cdot Q}{k} \tag{4.1}$$

with

$$\frac{1}{E_{\text{eff}}} = \frac{(1-\nu_1^2)}{E_1} + \frac{(1-\nu_2^2)}{E_2}. \tag{4.2}$$

Here, ε is a number between 1.9 and 2.4 [162], $\langle r \rangle$ is the time-averaged contact radius, E_{eff} the effective YOUNG'S modulus, E_1, ν_1 and E_2, ν_2 are the YOUNG'S modulus and POISSON'S ratio of the cantilever tip and the sample, respectively, Q the quality factor of resonance, and k the spring constant of the cantilever.

The obtained non-quantitative phase image is a color-coded map of the phase shift at different surface locations and, thus, provides information about the size and location of different materials – expressed by different moduli – across the sample surface. Fig. 4.5b clearly shows the surface distribution of hard clay nanoparticles (orange spots) of a NC15-gel embedded in the soft PNIPAAm polymer matrix (light blue color). The location of these particles at the surface is confirmed by the height image (light blue spots in Fig. 4.5a). A disordered random particle distribution over an area of $(1 \times 1)\,\mu\text{m}^2$ can be observed where most of the particles appear separated and exhibit an approximately circular or fragmentary geometry. The measured surface profiles of selected clay particles, indicated in Fig. 4.5b, reveal particle diameters between 30 nm and 40 nm (Fig. 4.6). Size deviations can be attributed to different

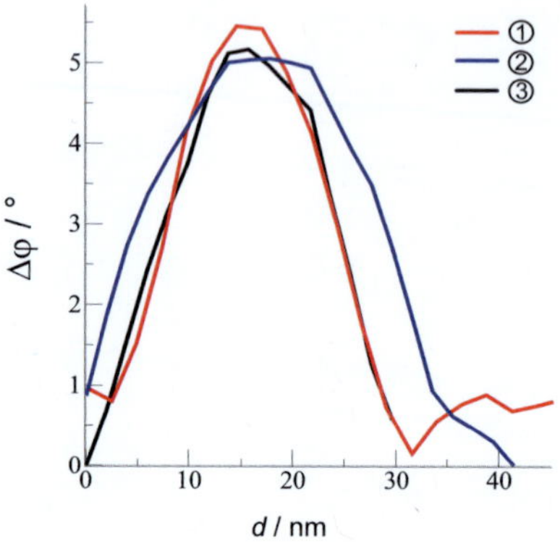

Figure 4.6: Surface profiles of clay nanoparticles at the spots indicated in Fig. 4.5b.

orientations of the clay particles at the surface. The dark blue regions are correlated to particles which are located well beneath the surface because the hardness and, thus, the phase shift changes due to the elasticity of the polymer surface layer covering the particles. Besides the existence of single separated particles, also particle agglomerations can be observed (white circles in Fig. 4.5b). The formation of these clay clusters may result from the relatively high clay concentrations leading to poor dispersion and consequently to an uneven particle distribution. The results obtained by AFM clearly confirm the SEM and TEM measurement results in [158].

4.2.4 Mechanical characterization

a. Uniaxial tensile measurements

To prepare the gel samples for the tensile tests, the pregel solutions with different clay concentrations (cp. Appendix A.2) were casted in tubular glass vessels and subsequently cross-linked by photopolymerization. As a result, cylindrical samples (diameter 5 mm, length 50 mm, initial cross section 19.6 mm^2) with low (NC6) and high (NC15) clay concentrations have been obtained. The samples were measured in their as-prepared hydrated state at 25 °C (set-up: Zwick-Roell, clamping length 10 mm, crosshead speed 50 mm/min). The obtained stress-strain curves for varying clay concentrations (Fig. 4.7a) show extraordinary mechanical properties with an elongation at break of about 1150 % for both NC6 and NC15 gels which is far above that of BIS-gels (Table 4.1). The BIS-gels were too brittle for mounting to the set-up and even broke during removal from the glass vessels. The photopolymerized NC-gels even show superior properties compared to thermally initiated NC-gels [163]. The very large strain values and tensile strengths of photoinitiated NC-gels can be attributed

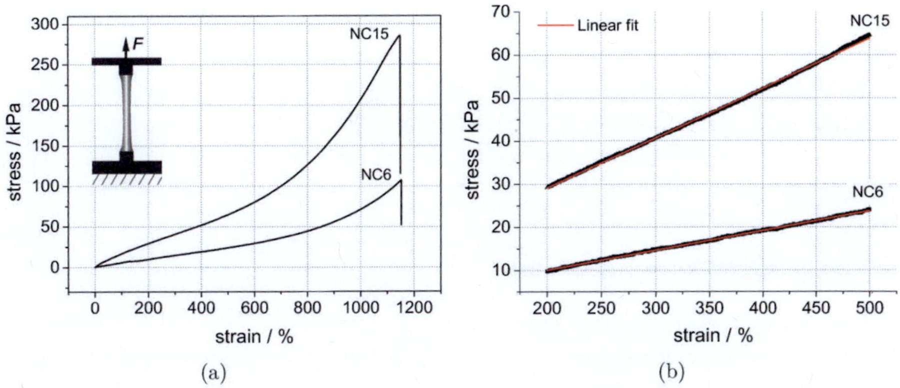

Figure 4.7: Stress-strain curve of photopolymerized clay nanocomposite hydrogels with varying clay content: (a) stress-strain curves (displacement-controlled, the stress is calculated using the constant initial cross-sectional sample area), (b) linear elastic range to determine the tensile moduli.

to a strong physical interaction between the polymer chains and the surface of the clay particles. During the polymerization reaction, the PNIPAAm chains adsorb to the clay surface and establish hydrogen bonding, i. e. the clay particles act as multifunctional crosslinks. In [153] it has been shown that the C=O and N-H groups of PNIPAAm are saturated by hydrogen bonds, where the involvement of C=O groups in the bonding process is more pronounced. There it was also assumed that large elongations lead to the rupture of polymer chains but the affinity of the PNIPAAm chains to the clay surface provoke new bonds at different locations such that a stable regenerative network may explain the outstanding mechanical properties.

From Fig. 4.7b it can be seen that different clay concentrations influence the elasticity of the material. NC6 gels show a lower tensile strength and tensile modulus (slope in Fig. 4.7b) compared to NC15 gels (Table 4.1). During the initial state of photopolymerization, the clay particles are covered by a PNIPAAm layer. For low clay concentrations, this layer is compact and only a small amount of remaining polymer chains connect to other clay particles leading to a soft network structure.

Table 4.1: Mechanical properties of Laponite-nanocomposite gels compared to conventional BIS-gels at 23 °C.

Sample	Tensile strength in kPa	Tensile modulus in kPa	Elongation at break in %
NC6	107	4.6	1148
NC15	284	11.6	1152
BIS-gels [164]	∼10	< 100	< 50

For increasing clay concentrations, the layer, surrounding the clay particles, becomes less compact because more clay sheets are available and, as a consequence, more PNIPAAm chains will connect to surrounding clay sheets. Thus, a more solid network structure with an elevated tensile modulus is formed [157]. However, even for high clay concentrations, the tensile moduli were found to be well below that of conventional BIS-gels.

b. Nanoindentation measurements

To directly compare the elasto-mechanical properties of NC- and BIS-gel sensing layers at different swelling states, force-distance curves have been recorded by nanoindentation (Nanowizard I from JPK Instruments mounted on an Axiovert 200M optical microscope (Carl Zeiss) together with a temperature-controlled liquid cell (AFM BioCell^{TM})) for both types of hydrogels at different temperatures. The set-up used allows measurements in an aqueous environment and simultaneous heating of the sample under precisely controlled conditions. The fabricated sensing layers (thickness 75 µm, cp. Section 4.2.2) have been attached to circular glass standard cover slips using an adhesive. The film thickness is sufficiently large that the indentation depth is much less than 10 % of the total film thickness and, thus, an interaction of the cantilever with the substrate can be excluded [165]. The glass coverslips with the attached hydrogel films were placed in the liquid cell which was filled with deionized water. Subsequently, the hydrogel layers were allowed to equilibrate for about 1 h.

The AFM cantilever (tip-less Si cantilever, CSC series, MicroMasch), modified with a glass sphere (Kisker Biotech, radius 4.75 µm), was vertically approached by

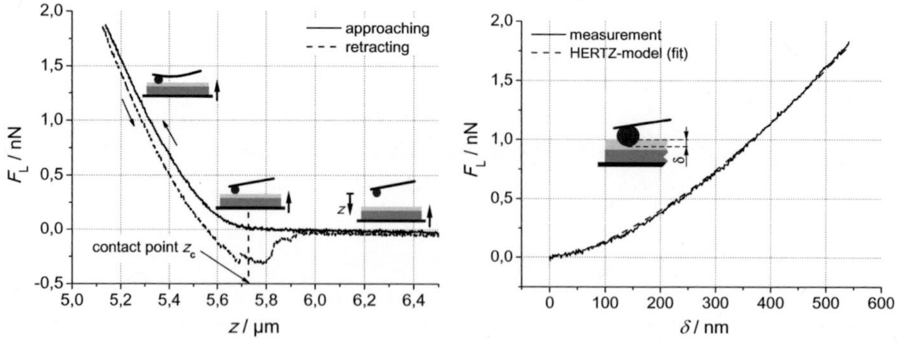

(a) (b)

Figure 4.8: Load applied to the cantilever as a function of (a) the piezo table position and (b) the indentation depth of the sphere into the hydrogel layer (NC15-gel at 30 °C). The HERTZ model was used to fit the experimental curve in (b).

the piezo table containing the sample until a contact between the sphere and the hydrogel surface is established ($z = z_c$) and the cantilever is deflected. Subsequently, the cantilever was retracted from the surface (Fig. 4.8a). For a hard material, the position z of the piezo table is proportional to the deflection d of the cantilever. For soft materials, the sphere on the cantilever penetrates the sample surface with an indentation depth δ. Thus, the deflection d of the cantilever at the same piezo position z is smaller for soft samples.

The values of the piezo probe-sample distance z and corresponding values of the cantilever deflection d were measured. From the cantilever deflection, the loading force F_L (force acting on the cantilever) can be calculated as $F_L = kd$ using the spring constant of the cantilever ($k = 66.2\,\mathrm{mN/m}$). The HERTZ-model predicts the applied load F_L on a sphere as a function of the indentation depth δ into a compliant half space as [166]

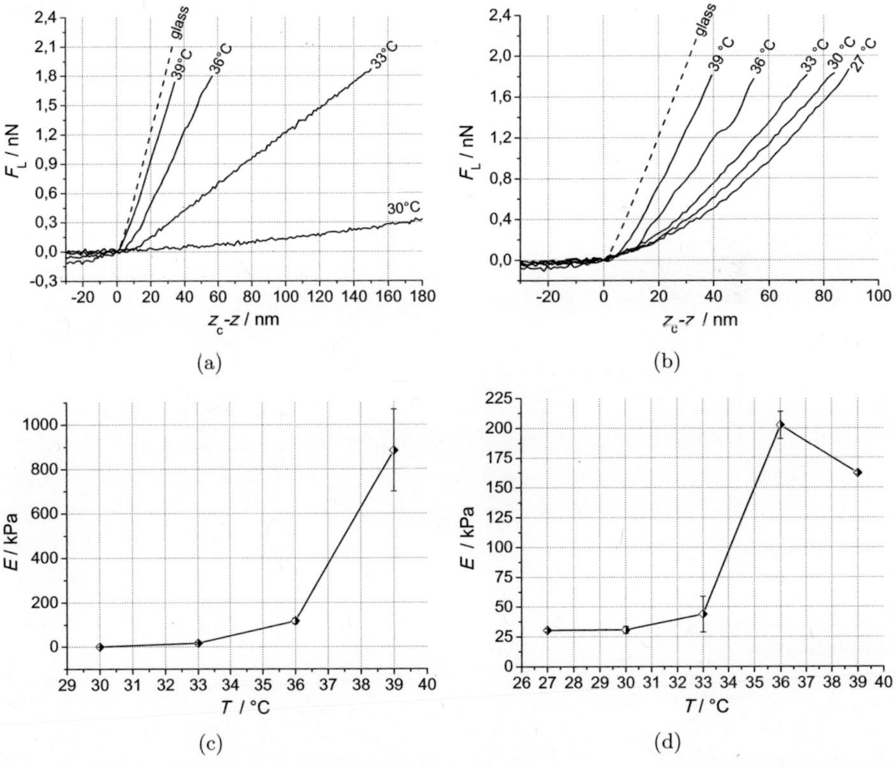

Figure 4.9: Force-distance curves for (a) NC15-gel, (b) BIS4-gel; temperature dependence of YOUNG'S modulus for (c) NC15-gel, (d) BIS4-gel.

$$F_{\mathrm{L}} = \frac{4}{3} \frac{E}{(1-\nu^2)} \sqrt{R} \delta^{3/2} \tag{4.3}$$

and can be fitted to the experimentally obtained $F_{\mathrm{L}}(\delta)$-curves (Fig. 4.8b) for estimation of YOUNG'S modulus E. Here, R and ν are the radius of the sphere mounted to the cantilever tip and the POISSON'S ratio of the hydroge layer, respectively. In this way, force-distance curves have been measured for NC15- and BIS4-gels for different temperatures (Figs. 4.9a,b). In order to verify reproducibility of the measurement data, five different locations on the same sample have been scanned five times sequentially for each temperature value. The values of the elastic modulus were obtained by adjusting the experimental curves to Eq. (4.3) (Figs. 4.9c,d).

With increasing temperature, it can be seen that the modulus values are also increasing. The moduli of the hydrogel layers are related to the swelling degree of the network. Below the LCST, the network chains are hydrated and exhibit flexible, expanded coil confirmation in water. Thus, in this swelling state, they are soft, resulting in small modulus values. The coil-globule transition above the phase transition temperature leads to dehydrated and collapsed network chains above the transition temperature. The tightly packed chains in globular confirmation show increased stiffness and, thus, significant higher modulus values. Although both curves show a similar qualitative behavior, the mechanical properties of both hydrogels differ which is especially apparent for $T = 30\,°\mathrm{C}$ and $T = 33\,°\mathrm{C}$ in the $F_{\mathrm{L}}(z_{\mathrm{c}} - z)$-curves (Figs. 4.9a,b). For NC15-gels below the LCST, the indentation depth is much higher compared to BIS4-gels, indicating small YOUNG'S moduli and, thus, a very soft network (Table 4.2). This qualitatively confirms the results obtained by tensile measurements. The quantitative deviations can be attributed to the different measurement methods and the different hydration states of the gels during measurement, i. e. tensile measurements have been performed on as-prepared samples, whereas the gel layers used for nanoindentation were fully hydrated.

Above the LCST, particularly for $T = 39\,°\mathrm{C}$, the NC15-gel shows a much higher modulus than the BIS4-gel. These results suggest that well above the phase transition temperature, NC-gels expel more water from the network compared to BIS-gels,

Table 4.2: Temperature-dependent YOUNG'S modulus of NC15- and BIS4-hydrogels measured by nanoindentation.

	E in kPa				
T	27 °C	30 °C	33 °C	36 °C	39 °C
NC15	–	0.5 ± 0.01	16 ± 3.4	116 ± 13	884 ± 185
BIS4	30 ± 1	31 ± 3	44 ± 15	202 ± 11	162 ± 1

leading to a more compact and stiffer network structure. The existence of tightly packed nanoparticles might also contribute to enhanced modulus values. The observed higher standard deviation may be a result of cluster formation of nanoparticles (cp. Section 4.2.3) due to the high crosslinking density and, thus, the existence of hard and soft regions across the sample surface. These results correspond to the observation of the mechanical properties of dry gel layers. Here, NC-gels show hard, glass-like properties and are brittle whereas BIS-gels exhibit flexible foil-like properties.

4.3 Sensor set-up

The sensor set-up to evaluate the sensor properties of NC-gels (Fig. 4.10) is well comparable to the one in Section 3.3.2. The dry hydrogel layers (cp. Section 4.2.2) are attached to a silicon support ($(0.9 \times 0.9)\,\mathrm{mm}^2$) using an adhesive and mounted on a modified TO8 package. The pressure sensor chip is also attached to the package by an adhesive such that the hydrogel is enclosed between the cavity of the pressure sensor chip and the silicon support. The fluidic channels enable continuous supply of the measuring solution by means of a syringe pump. For precise temperature control, all measurements have been carried out in a temperature chamber.

The attachment of the nanocomposite hydrogel layers to the silicon support by an adhesion promoter, as described for macroporous gels in Section 3.3.1, could not be successfully applied here. After swelling in water, the hydrogel layers pealed of from the substrate and showed a severely wrinkled surface. It is assumed that the high water uptake of clay nanocomposite gels leads to a decreased polymer volume fraction at equilibrium and, thus, to a high bidirectional swelling stress (cp. Eq. (3.4)). The reliable attachment of these gels to rigid substrates should by a major concern in future research.

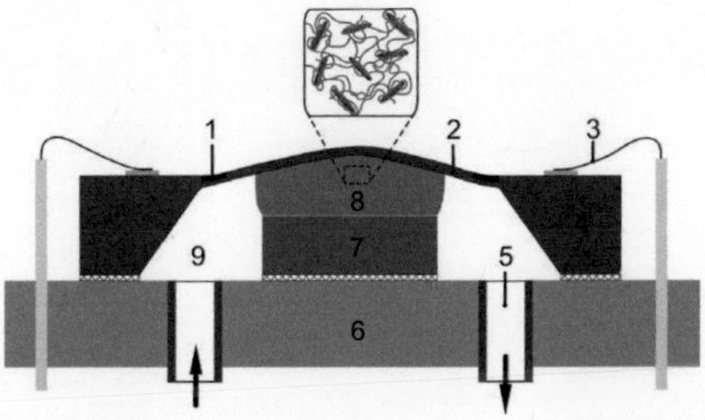

Figure 4.10: Cross-section of the sensor set-up for a sensor based on Laponite clay nanocomposite hydrogels. 1 piezoresistors, 2 flexure plate (thickness 20 µm), 3 electrical interconnect, 4 silicon pressure sensor chip ($(5.2 \times 5.2 \times 0.4)\,\mathrm{mm}^3$), 5 fluidic channels, 6 modified TO8 package, 7 silicon support, 8 PNIPAAm hydrogel layer, 9 measuring solution.

4.4 Sensor characterization

4.4.1 Sensor conditioning

After sensor assembly, the hydrogel is in its unswollen dry state. Prior to measurement, a conditioning process, i. e. the swelling of the dry hydrogel sensing layer in deionized water at 20 °C, is carried out. In this way, the sensor is brought into a stable reference state, assuring reliability of the subsequent measuring process [8]. The relative change in sensor output voltage for the BIS4-gel shows strictly increasing behavior with time but a three-step process for the NC-gels with different clay concentrations. This suggests the existence of simultaneous, superimposing processes for the initial swelling of NC-gels in water (Fig. 4.11a):

(I) For the initial fast swelling process, a clear dependence of the output signal amplitude on the clay concentration can be observed, i. e. a higher clay concentration leads to a higher water uptake of the gel and, thus, to a higher signal amplitude. This result is in contradiction with the predictions of the FLORY-REHNER theory where the osmotic swelling pressure is decreased by increasing crosslinking concentration (cp. Eq.(2.14)) but might be explained by the existence of an ionic contribution of the osmotic pressure due to the charged clay particles ($\Pi_\mathrm{m} + \Pi_\mathrm{ion}$). Moreover, since clay gels with low clay concentrations (e. g. NC6) are very soft (cp. Section 4.2.4), the restoring force of the flexure plate might impede swelling of the gel in vertical direction, leading to smaller output signal changes for lower clay concentrations.

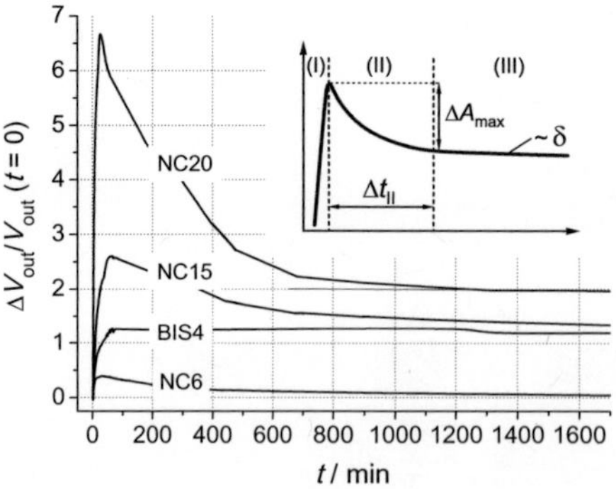

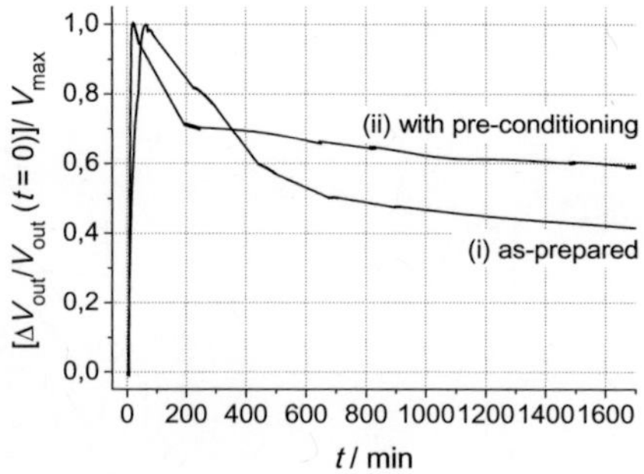

Figure 4.11: Characteristic change of the relative sensor output signal for sensors containing (a) conventional BIS4- and NC-gels with different clay concentrations, (b) as-prepared and pre-conditioned NC15-gels, respectively.

(II) The second step is characterized by a decreasing output signal where the signal drop ΔA_{max} is increasing with increasing clay concentration (Fig. 4.12b). A similar behavior has been observed by HARAGUCHI *et al.* for free swelling of nanocomposite PNIPAAm and PDMAA hydrogels cross-linked with Laponite XLG [167]. There, it has been assumed that, after initial water absorption of the network, the diffusion of sodium ions out of the gel and the possible substitution by hydrogen ions leads to a decrease of the ionic contribution of the osmotic pressure and, thus, to deswelling. For highly cross-linked gels (e. g. NC20), the number of sodium ions is high and the release of these ions is more time-consuming (Δt_{II}) and induces a larger decrease in hydrogel volume compared to lower cross-linking concentrations (Figs. 4.12a,b). Additionally, this trend might be assisted by the mechanical restoring force of the flexure plate acting on the soft hydrogel layer. To deconvolute both effects, the initial swelling curves of two sensors containing NC15 hydrogel layers in different conditioning states have been recorded:

- for a dry as-prepared hydrogel layer,
- for a pre-conditioned hydrogel layer. This means that, before mounting of the pressure sensor chip, the hydrogel layer attached to the support has been placed in deionized water, was allowed to swell for 72 h and was subsequently dried and the pressure sensor chip was mounted on the TO-8 package.

After supply of deionized water to both sensors, the curves in Fig. 4.11b have been obtained. Here, the drop in sensor signal of the sensor with pre-conditioned hydrogel layer is much less pronounced compared to the sensor containing the as-prepared gel layer and the non-linear decrease is completed faster. Thus, for the latter, the release

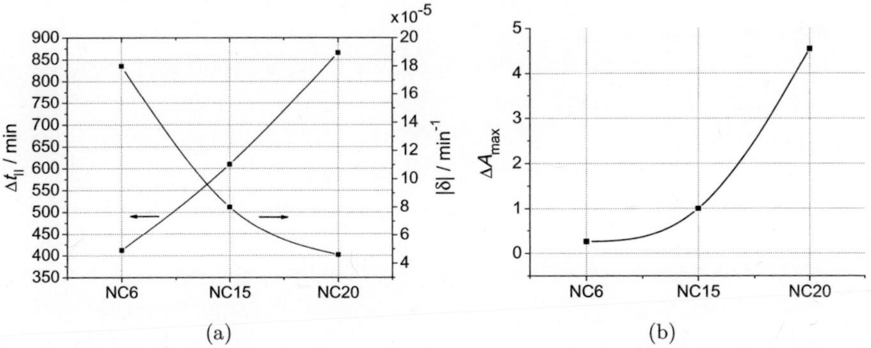

Figure 4.12: (a) Time span Δt_{II} for the non-linear decrease in sensor output signal (step II) and the absolute value of the linear drift rate $|\delta|$ (step III), (b) relative sensor output signal drop ΔA_{max} (step II) for different clay concentrations.

of sodium ions and the restoring force of the flexure plate lead to the slow decreasing signal. However, for the sensor containing the pre-conditioned hydrogel layer, all sodium ions were already released from the network during pre-conditioning and the observed signal decrease is merely attributed to the stiffness of the flexure plate.

(III) The third step is characterized by a slow linear decrease with rate δ. This linear drift effect can be attributed to the viscoelastic hydrogel properties, i. e. the vertical hydrogel extension is counteracted by the flexure plate and the gel evades horizontally within the sensor chip cavity. With increasing clay concentration, the drift rate decreases because the network becomes stiffer and the influence of the flexure plate is less pronounced (Fig. 4.12a).

From Fig. 4.11a one can conclude that the high swelling degree of NC-gels in water, commended in literature for free swelling gels [163], can be only applied to a limited extend for the enhancement of the signal amplitude of piezoresitive hydrogel sensors. For low cross-linking concentrations (e. g. NC6), the signal amplitude almost vanishes after step II and is far below that of conventional BIS-gels because these hydrogel layers are too compliant. Thus, only higher clay concentrations ($\geq$ NC15) lead to a proper signal amplitude which exceeds that of BIS-gels and show a stable signal with small drift rate in the steady state.

4.4.2 Temperature-induced sensor response

a. Phase transition

The temperature sensitivity of water-swollen NC-gels with different clay concentrations and BIS4-gels was studied by gradually increasing the temperature under precisely controlled conditions in a temperature chamber (Fig. 4.13a). It has been observed that the incorporation of clay particles into the PNIPAAm network does not elementary alter the sensor characteristics and a decrease of the sensor signal with increasing temperature is equally clearly visible as known for PNIPAAm gels. The complanate characteristics of the sensor with NC6-hydrogel layer is attributed to the very soft network structure as described in the previous section. For higher clay concentrations, the characteristic curves show almost the same behavior as for the BIS4-gel. However, a slight difference can be observed, i. e. a straightening of the curves with increasing clay concentration. This might be the result of altered water-polymer-clay interaction but needs to be studied further.

The first derivation of the characteristic curves gives information about the phase transition temperature (Fig. 4.13b). The local minima of these curves reveal a phase transition temperature of $T_c \approx 32\,^\circ\mathrm{C}$ for BIS4-, NC6-, and NC15-gels. The observed

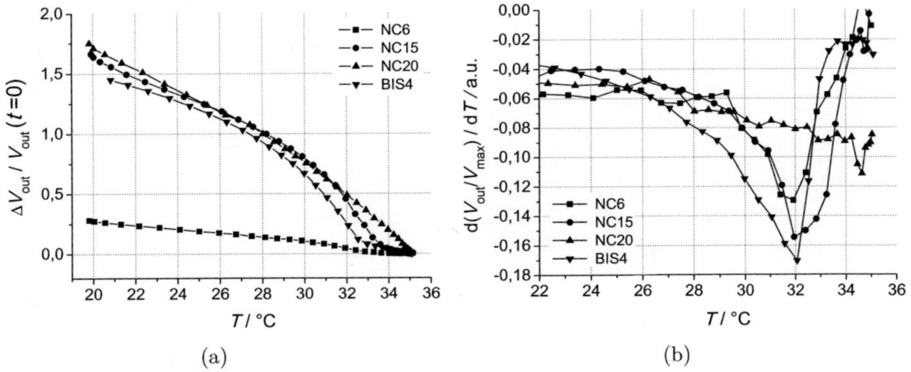

Figure 4.13: (a) Normalized sensor response as a function of temperature, (b) first derivative of the curves in (a) for determination of the phase transition temperature.

differences can be attributed to measurement uncertainties. Thus, the hydrophilic clay particles do not influence the sensitive hydrophilic-hydrophobic equilibrium of these gels in a wide range of clay concentration. These results confirm the determination of T_c by differential scanning calorimetry in [153] where a value of $T_c \approx 33\,°C$ was found for clay concentrations between NC0.5 and NC10. Interestingly, these observations do not apply to sensors with highly cross-linked NC20-gels. Here, no local minimum can be found for the observed temperature range. The apparent minimum at $T = 34.6\,°C$ is an artifact because even measurements up to 42 °C did not show a distinct minimum (data not shown). Hence, for very high clay concentrations near the dispersion limit of clay particles, it is assumed that a continuous phase transition rather than an abrupt one occurs and that no exact phase transition temperature exists.

b. Repeatability of the sensor signal

The application of repeating temperature cycles induces different hydration states of the gel layer and, thus, allows the analysis of the repeatability of the sensor signal. Here, cyclic temperature changes between 20 °C and 35 °C have been applied. For both conventional BIS4- and NC20-gels, a very good signal repeatability can be observed (Fig. 4.14). After the conditioning process (cp. Section 4.4.1), it is believed that the polymer chains of the NC-gel layer have adopted an optimum configuration such that a stable sensor response is achieved. For sensors with NC15 sensing layers, a similar repeatable, stable sensor signal has been observed. However, for sensors containing NC6-gel layers, a significant drift of the overall sensor signal with time and a distinct decrease of the signal amplitude from one cycle to the next was found.

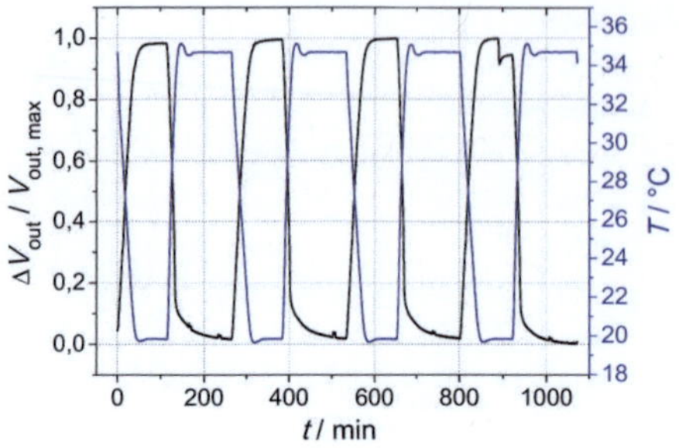

(a)

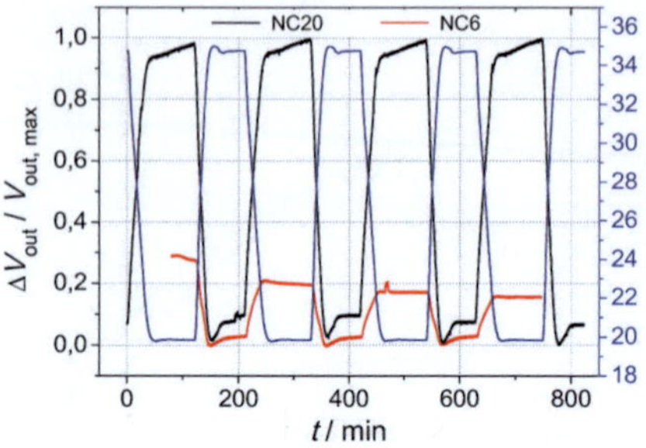

(b)

Figure 4.14: Normalized sensor response to cyclic temperature changes between 20 °C and 35 °C: (a) sensor with conventional BIS4-hydrogel layer, (b) sensors with NC6- and NC20-hydrogel layer, respectively. The sensor signal for NC6 is normalized to the output signal difference ΔV_{out} of the sensor with NC20-gel for comparability.

c. Time-dependent sensor response

The time-dependent sensor output signal for temperature excitation by a ramp signal has been measured to study the dynamic sensor response of NC-gels (Fig. 4.15). For decreasing temperature, the hydrophilicity of the gel network increases and water is absorbed, leading to an increase in sensor output signal. The dynamic sensor behavior for this swelling process is qualitatively the same for BIS- and NC-gels (Fig. 4.15a). Here, the time to reach equilibrium is determined by the time constant of the temperature control. For an immediate temperature step, the sensor is expected to react much faster. From Fig. 4.15a, a linear temporal drift of the sensor signal in quasi-equilibrium ($t > 40\,\text{min}$) can be observed for sensors containing NC-gels, whereas sensors with BIS-gel layer show nearly no drift. The drift rate increases with increasing clay concentration ($3.6 \times 10^{-4}\,\text{min}^{-1}$ for NC6 to $9.8 \times 10^{-4}\,\text{min}^{-1}$ for NC20). This behavior is well-known from ionic hydrogels such as PVA/PAA and can be attributed to the charged clay particles which repel each other when the hydrogel layer is in its soft hydrated state, leading to a slow linear network expansion.

A temperature increase leads to increasing hydrophobicity of the gel layers and the water inside the hydrogel matrix is continuously expelled from the network, leading to a decrease in sensor output signal (Fig. 4.15b). Here, a distinct difference of sensor dynamics for sensors using BIS-gels compared to sensors with NC-gels is apparent. Sensors with NC-gels show a much faster deswelling behavior and the sensor response even follows the overshoot of the temperature signal. However, sensors with BIS4-gels show very long deswelling times. For $t \leq 19\,\text{min}$, the sensor response of both sensors is determined by the rate of temperature increase. For $t \approx 19\,\text{min}$, the phase transition temperature is reached. For BIS4-gels above the phase transition temperature, an

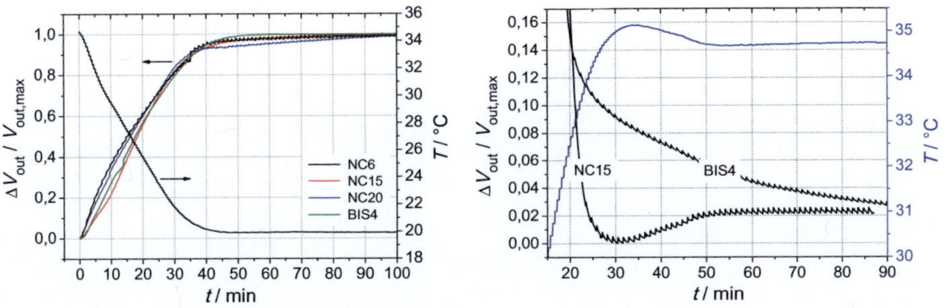

(a) (b)

Figure 4.15: Normalized sensor response to (a) decreasing and (b) increasing temperature ramps.

effect known as shrinkage barrier effect or skin effect occurs, i. e. the immediate formation of a dense polymer layer at the hydrogel surface. Due to thermal diffusion with a constant rate, the hydrogel surface is effected first by the temperature stimulus. As a result, a collapsed skin-like hydrophobic layer develops at the hydrogel surface, whereas the inner hydrogel volume is still in its swollen state. Thus, the remaining water inside the hydrogel is trapped and the release of this water into the surrounding solution can only take place with a reduced rate through the dense skin layer, leading to very long deswelling times. This phenomenon has been observed for BIS-gels by ^{1}H NMR imaging in [152]. Obviously, for NC-gels, this phenomenon does not occur. For $t > 19$ min, the response of the sensor containing the NC15-gel is not influenced by the exceedance of the phase transition temperature and the sensor signal directly follows the temperature excitation signal. This sensor behavior was found to be independent of the clay concentration and, hence, also occured for highly cross-linked NC20-gels. Thus, the formation of a dense polymer layer is suppressed and even for a collapsed network, an uniform solution transport into and out of the hydrogel is possible. This might be explained by two possible effects:(i) the reduced crosslinking density of NC-gels compared to BIS-gels leading to networks with a larger mesh size [168], (ii) the size of the clay particles may also have an effect, i. e. is is assumed that the density of the collapsed network might be limited by the diameter of the clay particles such that channels for a uniform solution transport are available.

4.4.3 Sensor response to organic co-solvents

a. Transfer characteristics

To study the sensor response to organic co-solvents, ethanol was chosen as a model solvent. The response to other solvents like methanol, 1-propanol, or 2-methyl-2-propanol is comparable [169]. In particular, ethanol-water mixtures with different

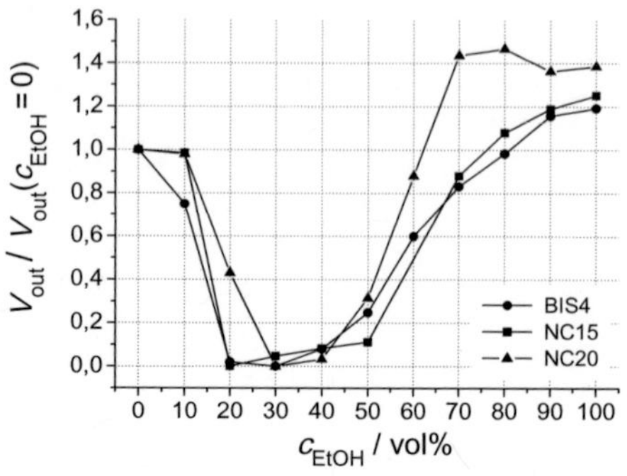

Figure 4.16: Isothermal sensor transfer characteristics for sensors with BIS4-gel and NC-gels with different clay concentrations. The sensor output voltage is measured at 23 °C and normalized to the output voltage for pure water ($c_{\mathrm{EtOH}} = 0$) to enable comparability.

ethanol concentrations have been applied to the sensor. The isothermal sensor transfer characteristics shows reentrant transition behavior as known for sensors based on PNIPAAm hydrogels [8] and is comparable for conventional BIS- and NC-gels (Fig. 4.16). Similar to the observations for the temperature-induced sensor response, the characteristic curves for sensors with BIS4- and NC15-gels are almost identical. For both only a very small sensitivity in the range 20 vol%< c_{EtOH} <50 vol% can be observed. For sensors containing NC-gels with high clay concentration (NC20), the characteristic curve appears compressed. This suggests that for high clay concentrations, the sensitive hydrophilic/hydrophobic equilibrium, determined by the polymer-solvent interaction, is additionally influenced by the hydrophilic clay particles. This reflects in both flanks of the characteristic curve: (i) the deswelling process is not finished until $c_{\mathrm{EtOH}} = 30$ vol% is reached and (ii) re-swelling can be already observed for $c_{\mathrm{EtOH}} > 40$ vol% with a steeper slope, indicating a higher solvent uptake. For $c_{\mathrm{EtOH}} \geq 70$ vol%, the curve reaches a plateau, suggesting that either the maximum capacity of solvent absorption has been reached or that the stiffness of the flexure plate suppresses further swelling of the soft hydrogel layer.

b. Time-dependent sensor response

Ethanol concentrations below 50 vol% lead to water release of the hydrogel layer and, thus, to deswelling, resulting in a decrease in sensor output signal. For conventional BIS4-gels, the contact of the proceeding solvent front with the gel surface leads to deswelling of the outer gel layer whereas the inner gel is still fully swollen. Similar to the observations during temperature stimulation, a dense polymer layer acting as shrinking barrier is formed and significantly slows down the release of water into the surrounding solution. This can be clearly seen by the time-dependent sensor response

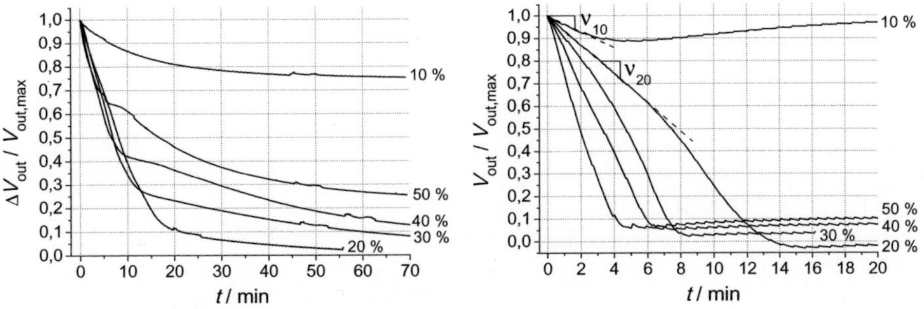

(a) (b)

Figure 4.17: Normalized time-dependent sensor output voltage at 23 °C: sensor with (a) BIS4-hydrogel layer, (b) NC15-hydrogel layer.

in Fig. 4.17a where a fast, almost linear signal decrease is followed by a very slow asymptotic signal behavior. Here, the dynamic sensor response is determined by the concentration gradient between the solution and the water-swollen gel but also by the formation of a heterogeneous gel structure during the deswelling process.

For sensors with NC15-gel layer (Fig. 4.17b), the long deswelling phase does not occur and only a fast signal decrease, immediately followed by the attainment of the steady state, can be observed. Thus, these sensors react much faster in the observed concentration range. Furthermore, the different concentration gradients between the solution and the inside of the hydrogel result in different deswelling characteristics accompanied by different slopes

$$\nu = k \cdot (c^{\mathrm{gel}}_{\mathrm{EtOH}} - c^{\mathrm{ext}}_{\mathrm{EtOH}}), \tag{4.4}$$

where k is the release rate of water into the surrounding solution, and $c^{\mathrm{gel}}_{\mathrm{EtOH}}$, $c^{\mathrm{ext}}_{\mathrm{EtOH}}$ are the ethanol concentration inside the gel and in the surrounding solution, respectively. Since the gel layer was pre-swollen in water ($c^{\mathrm{gel}}_{\mathrm{EtOH}} = 0$) Eq. (4.4) simplifies to

$$\nu = -k \cdot c^{\mathrm{ext}}_{\mathrm{EtOH}}. \tag{4.5}$$

Thus, the slope of the linear decrease in the first minutes of the measurement process can be used for determination of the concentration (Fig. 4.18). This enables the detection of solvent concentration between 20 vol% and 50 vol% within a few minutes which is not possible with sensors using conventional BIS-gels. The time-dependent sensor behavior thus only depends on the concentration gradient whereas the formation of a heterogeneous network structure by a dense barrier layer does not occur. This can be confirmed by measuring the optical transparency of the network for different ethanol concentrations (Fig. 4.19). BIS-gels as well as NC-gels, which

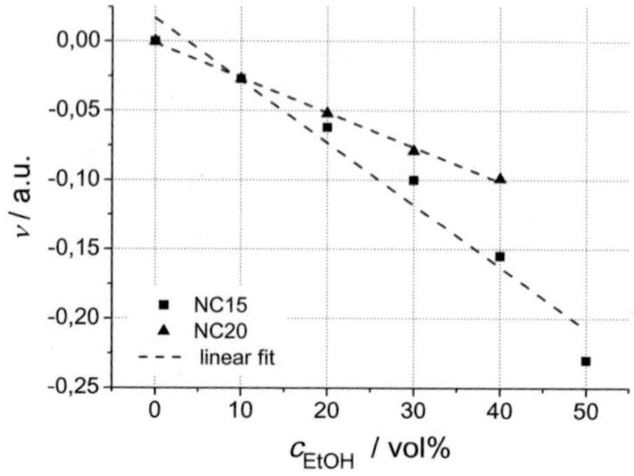

Figure 4.18: Slope of the linear region of the sensor output signal in the first 240 s as a function of the ethanol concentration in the surrounding solution for sensors with NC15 and NC20 hydrogel layers.

are swollen in water, are transparent for visible light. For BIS-gels, the formation of the barrier layer for low ethanol concentrations leads to a phase separation and, thus, to a heterogeneous network structure reflecting in an opaque appearance of the gel and in a drop in optical transmission. For NC-gels, only a slight decrease in optical transmission can be observed, revealing the maintainance of a transparent network. Hence, it can be assumed that no significant phase separation occurs and that a homogeneous network structure exists. This favors a uniform solution transport which is not impeded by the network structure itself.

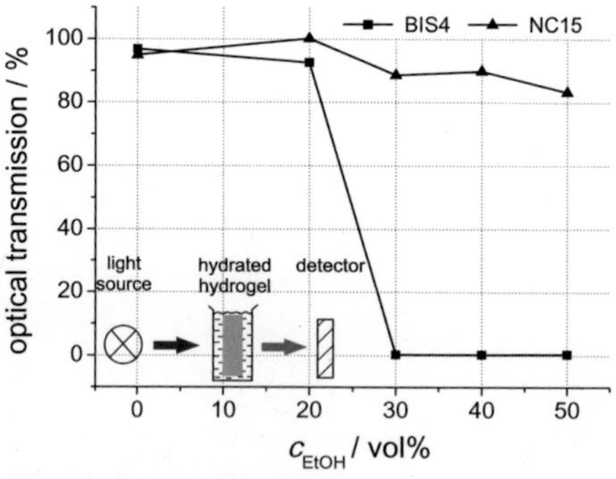

Figure 4.19: Optical transmission ($\lambda = 550\,\text{nm}$) of BIS4- and NC15-gel layers depending on the ethanol concentration.

4.4.4 Sensor response to monovalent salt ions

The addition of small salt concentrations to a water-swollen PNIPAAm-hydrogel influences the hydrogen bonds between water and polymer and, consequently, effects the phase transition temperature and the swelling state of the gel. The LCST shift particularly depends on the position of the involved anions in the HOFMEISTER series and is thus ion-specific, following the order $SCN^- < I^- < NO_3^- \approx Br^- < ClO_4^- < Cl^- < F^- < H_2PO_4^- < S_2O_3^{2-} < SO_4^{2-} < CO_3^{2-}$ [170]. Here, aqueous NaCl and NaI salt solutions containing both cosmotropic (Cl^-) and chaotropic (I^-) anions have been used as model solutions to study the sensor response. For conventional BIS-gels, the sensor response to NaCl is characterized by a monotonically decreasing signal due to hydrogel shrinking. It is induced by the "salting-out" effect of PNIPAAm (Fig. 4.20) [8]. The Cl^- anions polarize water molecules attached to the amide group of PNIPAAm by hydrogen bonding, thereby weakening this interaction and leading to salting-out. On the contrary, for small NaI concentrations, a "salting-in" effect can be observed leading to an increasing sensor output signal. ZHANG *et al.* suggested that the direct binding of I^- ions to the amide group are responsible for the salting-in effect [170–172]. However, very recent results obtained by simulation claim that the affinity of I^- ions to the hydrophobic part of PNIPAAm is responsible for this effect and, as a consequence, the amide groups are exposed to the solvent [173]. This in turn leads to a pronounced interaction between sodium cations and the oxygen of the carbonyl group.

Interestingly, sensors based on NC-gels show a completely different behavior. Here, salting-in can be observed for low NaCl concentrations. The concentration range where this effect occurs depends on the clay content (Fig. 4.21a). Salting-in can be observed for concentrations of $c \leq 0,01$ mol/l and $c \leq 0,1$ mol/l for NC15- and NC20-gels, respectively. Above these concentrations, salting-out leads to a decrease

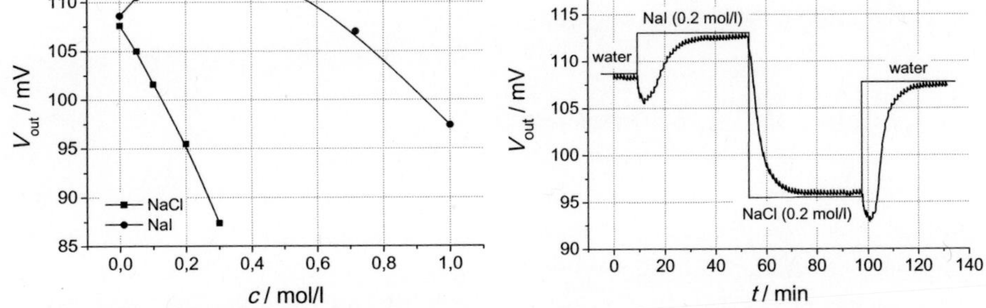

(a) (b)

Figure 4.20: Sensor response to salt concentrations of BIS6-PNIPAAm-based sensors: (a) sensor transfer characteristics, (b) time-dependent sensor signal for subsequent supply of 0.2 mol/l NaI and NaCl measuring solution.

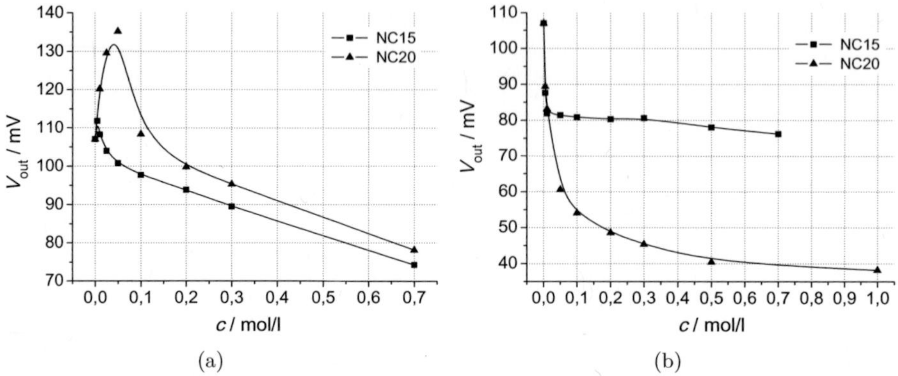

Figure 4.21: Sensor transfer characteristics depending on salt concentrations of NC-PNIPAAm-based sensors for (a) NaCl and (b) NaI.

in sensor output signal below that of pure water. For measurement of NaI, the sensor output signal is solely determined by the salting-out effect and, thus, shows a monotonic decreasing behavior. The observed differences in the sensor properties between sensors with BIS-gels and NC-gels might by attributed to the interaction of the amide groups with the clay particles forming strong hydrogen bonds as shown by IR-spectroscopy in [153]. In this way, the ionic species in solution cannot interact with the oxygen and nitrogen atoms of the amide groups but might interact with the hydrophilic clay surface. For BIS-gels, these groups are free to interact with water molecules and ions. Furthermore, for BIS-gels, salting-out expresses in a linear decreasing sensor signal whereas for NC-gels, a non-linear decrease can be observed. This might be attributed to the ionic character of NC-gels due to the charged clay particles. The exact water-ion-polymer-clay interaction for different ionic species is not trivial and should be investigated in further studies.

4.5 Conclusions and summary

In this chapter, for the very first time, photopolymerizable hybrid organic/inorganic hydrogels based on PNIPAAm cross-linked with different concentrations of Laponite XLS nanoparticles have been studied with respect to applications in piezoresistive microsensors. Structural characterization by EDX and AFM showed the successful formation of sub-100 µm hybrid sensing layers by photopolymerization where inorganic Laponite particles are randomly distributed in the organic PNIPAAm matrix.

Mechanical characterization revealed sensing materials with extraordinary toughness, very high elongation at break and tensile moduli, far exceeding that of conventional BIS-gels. In particular, AFM nanoindentation measurements showed that, despite the superior mechanical properties, thin hydrated sensing layers below the LCST are very soft compared to conventional BIS-gels but show elevated YOUNG'S moduli well above the LCST. Thus, at low temperatures, the compliance of the flexure plate shows a distinct influence on the gel layer, i. e. the swollen sensing layer acting on the flexure plate is pushed back by it, leading to a very small or even vanishing sensor signal and to drift effects. This behavior was found to be most pronounced for low clay concentrations. Therefore, only gels with high clay concentrations $\geq$ NC15 are suitable for sensor applications. The high swelling capacity known for free swelling NC-gels can thus only be used to a very limited extend for piezoresistive microsensors. Sensors containing NC15- and NC20-hydrogel layers show only a very small drift of the sensor signal after the mandatory conditioning process and very good signal repeatability.

Temperature stimulation showed that the time constant for deswelling can be greatly reduced by using NC-gels. The heterogeneous network formation (shrinkage barrier effect) observed for BIS-gels, which leads to a very slow sensor response, does not exist for NC-gels. This was also found for ethanol concentrations below 50 vol% and was confirmed by optical transmission measurements. The slope of the time-dependent sensor signal can be advantageously used for determination of ethanol concentrations within a few minutes. For temperature stimulation and ethanol measurements, it can be concluded that the influence of incorporated clay particles on the equilibrium response of the sensor is marginal and just appears for very high clay concentrations close to the dispersion limit. However, the time-dependent sensor response is primarily effected by the hybrid network structure which enables a uniform solution transport and leads to an improved sensor response.

In contrast to BIS-gels, a salting-in effect has been found to determine the sensor response for small NaCl concentrations and salting-out has been found for NaI for clay nanocomposite hydrogels in the observed concentration range. This suggests that the clay particles play an important role by influencing the water-ion-polymer equilibrium. The exact interaction is still unclear and should be studied in detail.

III Hydrogel sensors with force compensation

5 Force compensation method [1]

In this chapter, the compensation method is described with particular respect to sensor technology. The description is based on linear system theory and builds up the theoretical foundation for Chapter 6 where the set-up and implementation of a pressure-compensated hydrogel-based sensors is discussed. The goal of this chapter is a general understanding of the system behavior of a closed-loop sensor and to draw specific conclusions for sensor design and properties towards realization of high sensitivity, suppression of static systematic deviations, fast response, independence of interfering and long-term drift effects like material drift/relaxation and aging, as present in conventional open-loop hydrogel chemical sensors. The influence of three different linear standard controller-types, i. e. Proportional (P), Proportional-Integral (PI), and Proportional-Integral-Derivative (PID), on the sensor behavior is studied. Conclusions are drawn which controller type should be applied for realizing specific sensor properties. In Section 5.1 the basic concept of compensation is introduced, whereas in Section 5.2 a general sensor model is presented and different transfer functions, which mainly cover the transfer characteristics of current sensors, are proposed. The modified system structure, which results from the feedback for compensation, is shown and the resulting transfer functions are calculated (Section 5.3). These transfer functions are systematically analyzed with respect to certain sensor properties mentioned above.

5.1 Basic concept of compensation

For sensors using the deflection method, the sensor output signal is a direct measure of the input signal. This method comprises the lowest demand in terms of system complexity. However, the sensor output signal is directly determined by the sensor properties, leading to a pre-determined, but often insufficient sensor performance for specific applications. A different method known in measurement technology is the method of compensation. Possibly the most prominent example is the mechanical beam balance [175], where the beam angle is altered by the addition of a sample weight into one tray and subsequently restored by adding of the same weight to the other tray. This results in a balanced null-deflection state of the beam. The

[1] Considerable parts of this chapter are based on a publication by the author in the "*Journal of Sensors and Sensor Systems*" [174].

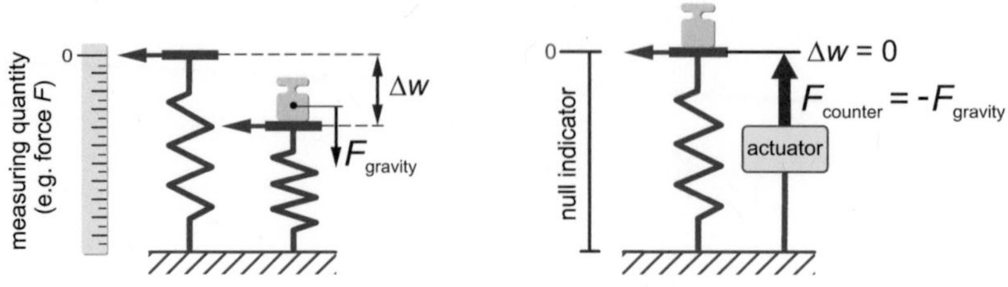

(a) (b)

Figure 5.1: Simplified schematic illustration of the principle of operation of (a) deflection method and (b) compensation method. The spring represents an arbitrary transducer, whereas the mass of the weight generally represents the input signal.

counterweight can be determined more precisely than the beam angle, which leads to an overall higher measurement precision compared to the deflection method. In general, the compensation method can be applied to a wide range of sensors, almost irrespective of the sensor principle. However, the applicability of this method implies the existence of an actuator unit (Fig. 5.1) which is able to physically generate the compensation quantity. In this method, the actual sensor output signal is measured, subtracted from a reference value, and the difference is fed back to the sensor input via a controller/actuator unit until the difference between output signal and reference value reaches zero. Thus, the static sensor output signal remains constant for changes in the input signal. Nevertheless, the "force" provided by the actuator which is needed to maintain this balanced state is directly related to the input signal. Compared to the deflection method, which is an open-loop configuration, a sensor operated with the compensation method is a closed-loop system. This leads to different system

Table 5.1: Elementary comparison between deflection method and compensation method in measurement.

	Deflection method	Compensation method
System model	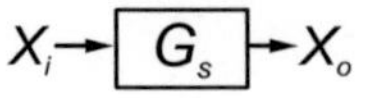	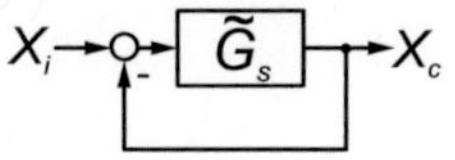
Input	Measurand X_i (e. g. force F)	Measurand X_i
Output	signal X_o (e. g. output voltage)	compensation quantity X_c (e. g. force)
Sensitivity	$S^\mathrm{defl} = \left\lvert \frac{X_\mathrm{o}}{X_\mathrm{i}} \right\rvert = \lvert G_\mathrm{s} \rvert$	$S^\mathrm{comp} = \left\lvert \frac{X_\mathrm{c}}{X_\mathrm{i}} \right\rvert = \left\lvert \frac{\tilde{G}_\mathrm{s}}{1+\tilde{G}_\mathrm{s}} \right\rvert$

properties solely due to the feedback structure as outlined in Table 5.1. Furthermore, because of the perpetuation of the balanced "undeflected" state, the sensor properties negligibly contribute to the overall system behavior leading to smaller measurement uncertainties and, hence, to better measurement results. Additionally, the properties of the feedback loop can be set and tuned systematically. This is the reason why the compensation method is used in particular in technical applications with high-precision requirements or expanded operation fields, like precision balances [176], AFMs (in constant force mode) [177], broadband lambda probes [178], hot-wire anemometers (in constant temperature operation) [179,180], MEMS-based accelerometers [181,182], and continuous non-invasive blood pressure monitoring [183]. As can be seen from these examples, the compensation method represents a general approach which can be applied to sensors from different domains, i. e. mechanical, physical, chemical, etc.

5.2 General sensor model for deflection method

5.2.1 Sensor model

To keep generality, we regard the sensor as a single-input single-output (SISO) system with input x_i and output x_o. The transfer behavior is described by the impulse response g_s with $x_\mathrm{o}(t) = g_\mathrm{s} * x_\mathrm{i}(t)$, irrespective of any specific sensor principle, as shown in Fig. 5.2. Here, x_i is the measurand and can be any physical, chemical or biological quantity, whereas x_o is assumed to be an electrical output signal (voltage or current) according to the general sensor definition, and $*$ is the convolution operator. Furthermore, we assume the sensor to be a linear time-invariant (LTI) system. The linear dynamic input-output relation of the sensor in the time domain can be completely described by an ordinary linear differential equation of order n ($n \geq m$; $n, m \in \mathbb{N}$) with constant coefficients in the following form:

$$\begin{aligned} &a_n \frac{\mathrm{d}^n x_\mathrm{o}}{\mathrm{d}t^n} + a_{n-1} \frac{\mathrm{d}^{n-1} x_\mathrm{o}}{\mathrm{d}t^{n-1}} + \cdots + a_1 \frac{\mathrm{d}x_\mathrm{o}}{\mathrm{d}t} + a_0 x_\mathrm{o} = \\ &b_m \frac{\mathrm{d}^m x_\mathrm{i}}{\mathrm{d}t^m} + b_{m-1} \frac{\mathrm{d}^{m-1} x_\mathrm{i}}{\mathrm{d}t^{m-1}} + \cdots + b_1 \frac{\mathrm{d}x_\mathrm{i}}{\mathrm{d}t} + b_0 x_\mathrm{i}. \end{aligned} \tag{5.1}$$

Here, $a_0, \ldots, a_n$ and $b_0, \ldots, b_m$ are constant coefficients which are explicitly determined by the sensor properties. However, in consideration of real sensor systems, basically three instrument types, which mainly cover the transfer characteristics of most sensors, can be classified according to their dynamic response as: (I) first-order systems (PT_1), (II) second-order systems (PT_2), and (III) differentiating second-order systems (PT_2-D). Thus, Eq. (5.1) can be simplified for these three cases. The resulting input-output relations in the time domain are given in Table 5.2. However, to obtain

Figure 5.2: System model of a sensor: (a) in the time domain, (b) in the frequency domain.

the output characteristics for a given input signal, the description in the time domain requires the solution of the differential equations. This may result in complicated solution algorithms, especially for more complex systems in closed-loop configuration later on. Moreover, the quantitative interpretation of the solution and the extraction of practical conclusions may be not straightforward due to bulky mathematical expressions. Therefore, the sensor transfer characteristics is transformed from the time domain to the frequency domain, which enables a powerful method for theoretical system description [184]. Thus, the sensor can be completely described by its transfer function

$$G_s(s) = \frac{X_o(s)}{X_i(s)} = \mathcal{L}\{g_s(t)\} = \int_0^\infty g_s(t)e^{-st}\mathrm{d}t. \tag{5.2}$$

Here, $\mathcal{L}$ is the LAPLACE operator and s the complex variable. The sensor transfer functions according to Eq. (5.2) for the three basic sensor types are given in Table 5.2. The general sensor coefficients in the input-output relations are substituted by the static sensitivity K and the characteristic time constants T, T_1, T_2, and T_3 of the respective system.

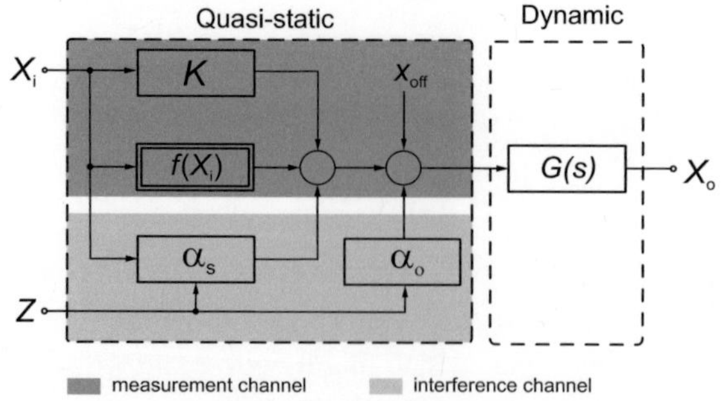

Figure 5.3: General sensor model with measurand X_i and sensor output X_o sub-divided in a quasi-static and a dynamic part and comprising a measurement and an interference channel. K intrinsic ideal sensitivity, $G(s)$ normalized sensor transfer function, α_s sensitivity coefficient, Z interference quantity, f nonlinearity in the transfer characteristics, x_{off} sensor offset, α_o offset coefficient.

Table 5.2: Sensor classification and assignment of the respective input-output relations and resulting transfer functions as well as common examples for the respective sensor types.

	Sensor type	Input-output relation $x_o(t) = g_s * x_i(t)$	Sensor transfer function $G_s(s) = X_o(s)/X_i(s)$	Examples
I:	First-order system (PT_1)	$x_o = \frac{\frac{b_0}{a_0}}{\frac{a_1}{a_0}\frac{d}{dt} + 1} \cdot x_i$	$\frac{K}{sT + 1}$	Temperature sensors [185]
II:	Second-order system (PT_2)	$x_o = \frac{\frac{b_0}{a_0}}{\frac{a_2}{a_0}\frac{d^2}{dt^2} + \frac{a_1}{a_0}\frac{d}{dt} + 1} \cdot x_i$	$\frac{K}{T_1^2 s^2 + T_2 s + 1}$	Velocity sensors, accelerometers [185]
III:	Differentiating second-order system (PT_2-D)	$x_o = \frac{a_0 b_1 \frac{d}{dt}}{\frac{a_2}{a_0}\frac{d^2}{dt^2} + \frac{a_1}{a_0}\frac{d}{dt} + 1} \cdot x_i$	$\frac{K T_3 s}{T_1^2 s^2 + T_2 s + 1}$	Piezoelectric sensors [186], pyroelectric sensors [187]

From Table 5.2 it is apparent that the sensor can be basically described by a quasi-static part and a time-dependent dynamic part as illustrated in the general sensor model in Fig. 5.3. The sensor model comprises the influence of the measurand X_i (measurement channel) and an interference quantity Z (interference channel). The static sensor behavior can vary from ideal linear behavior K due to the existence of non-linearities, which are apparent in most sensor applications. Both the linear part K and the non-linear part f directly depend on the measurand X_i. The static sensitivity is also changed by the interfering quantity Z, e. g. temperature in many sensor applications. When no input signal is applied ($X_i = 0$), only the static sensor offset x_{off} is measured as output signal. The sensor offset is independent of the input signal and, thus, does not contain information from the measurand. However, the offset is superimposed by Z and, hence, changed by the interference quantity. In most sensors, x_{off} is primarily altered due to temperature changes. The static sensitivity, the offset, and the influence of the interference quantity are additively superimposed and, together with the dynamic sensor part, are generating the sensor output signal X_o. Note that the order of the static and dynamic sensor part is arbitrary and could be also changed.

The sensor behavior can be expressed in a general sensor model as

$$X_o = [(K + \Delta K) \cdot X_i + (x_{off} + \Delta x_{off})] \cdot G(s), \tag{5.3}$$

with

$$\Delta K = f + Z \cdot \alpha_\mathrm{s} \tag{5.4}$$

and

$$\Delta x_\mathrm{off} = Z \cdot \alpha_\mathrm{o}. \tag{5.5}$$

It has to be pointed out that the influence of noise is neglected in this model. Based on this sensor model, the effect of the compensation method applied to such a sensor regarding the influence of non-linearities, interference quantities, and static sensitivity changes on the sensor output signal are of interest (see Section 5.3).

5.2.2 Quasi-static sensor behavior

For the description of the quasi-static sensor behavior, we assume that all dynamic processes are terminated and that the system is in an equilibrium state ($G(s) = 1$ for $s \to 0$). Hence, from Fig. 5.3 and Eq. (5.3) it becomes clear that the static sensor output signal is basically determined by two parts. One is the static sensor offset $\tilde{x}_\mathrm{off} = (x_\mathrm{off} + \Delta x_\mathrm{off}) = (x_\mathrm{off} + Z \cdot \alpha_\mathrm{o})$, which is determined by the intrinsic sensor offset x_off and the interference quantity Z, but independent of the input signal X_i. The other part is the sensitivity, which is usually altered by non-linearities and the interference quantity Z as a function of the input signal. According to Eq. (5.3), the resulting sensitivity can be expressed as the superposition of an ideal constant sensitivity K and a deviation ΔK as $\tilde{K} = (K + \Delta K)$. Here, again two cases have to be considered: (i) a time-independent sensitivity change (ΔK = constant) and (ii) a time-dependent sensitivity change (drift) ($\Delta K = \Delta K(s)$). For a time-independent sensitivity change, the interference quantity is constant over time. If one considers small time-independent deviations ΔK from the ideal behavior, the impact of relative systematic deviations ΔK on the static sensor output signal can be calculated using a TAYLOR-series and expressed as a systematic static deviation $\varepsilon_s^\mathrm{defl}$ as

$$\varepsilon_\mathrm{s}^\mathrm{defl} = \frac{\Delta X_\mathrm{o}^{\Delta \mathrm{K,s}}}{X_\mathrm{o}} = \frac{1}{X_\mathrm{o}} \frac{\partial X_\mathrm{o}}{\partial K} \cdot \Delta K = \frac{\Delta K}{K} \tag{5.6}$$

and is equal to the relative sensitivity change.

The second case of a time-dependent sensor sensitivity (drift) is apparent in many sensor applications and results from time-dependent variations of the interference signal. Here, this phenomenon is described by a simple model – known from mechanical sensors – which demonstrates the basic idea but keeps simplicity at the same time. The model originates from a simple spring-dashpot arrangement connected in series (MAXWELL-model) as illustrated in Fig. 5.4.

This mechanical model represents relaxation effects in spring elements under load

due to viscoelastic material properties. The resulting transfer function is used to describe linear drift effects irrespectively of any specific sensor principle.
In the frequency domain, $\tilde{K}$ can generally be described as

$$\tilde{K}(s) = K \cdot \frac{s+\delta}{s}, \tag{5.7}$$

where δ is a variable drift rate. Although we consider the static sensor behavior in this section, the phenomenon of drift is a dynamic process occurring in the quasi-static part of the sensor model. Therefore, we can illustrate the time-dependent sensitivity by calculating the sensor response due to a step change of the input signal X_i in the time domain which leads to

$$x_\mathrm{o}(t) = \mathcal{L}^{-1}\left\{\tilde{K}(s) \cdot \frac{1}{s}\right\} = K(1+\delta t). \tag{5.8}$$

The normalized step response x_o/K for the model in Eq. (5.8) for different drift rates δ is shown in Fig. 5.5. For a constant input signal, the static sensor output signal should ideally be constant. However, due to the drift phenomenon, the output signal increases with time. Thus, an explicit correlation between the input and the output signal is not possible without additional stipulations, i. e. the knowledge of the drift rate and the establishment of a constant measurement time.
For a time-dependent sensitivity change, the ideal sensitivity K has to be replaced by $\tilde{K}(s)$ from Eq. (5.7). Thus, the relative deviation of the output signal for a drift-affected sensor according to Eq. (5.6) becomes

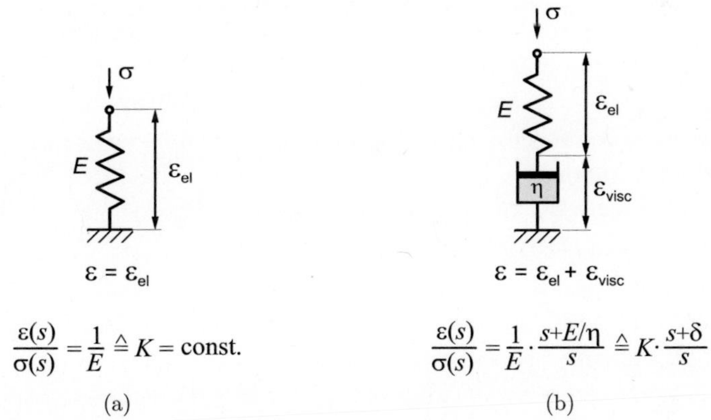

Figure 5.4: Mechanical analog and resulting transfer function of a sensor with (a) ideal linear time-independent sensitivity K (no drift) and (b) time-dependent sensitivity change due to viscoelastic material properties (drift); σ mechanical stress, ε mechanical strain, E YOUNG's modulus of the elastic spring, η viscosity of the dashpot.

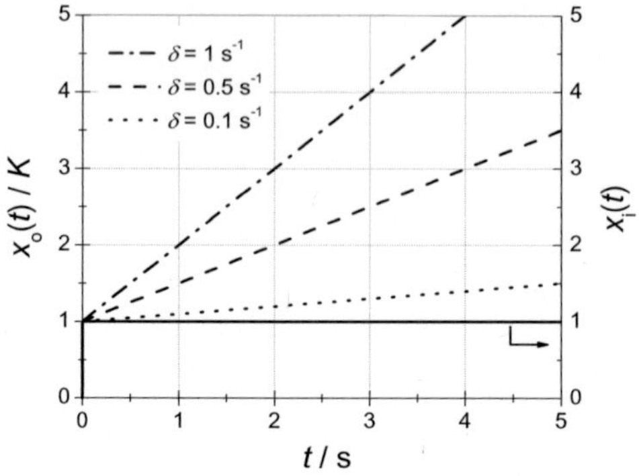

Figure 5.5: Normalized sensor output signal x_{o}/K (interrupted lines) for a unit step input (solid line) when considering a time-dependent sensor sensitivity (drift) with different drift rates δ.

$$\varepsilon_{\mathrm{d}}^{\mathrm{defl}} = \frac{\Delta X_{\mathrm{o}}^{\Delta\mathrm{K,d}}}{X_{\mathrm{o}}} = \left(1 + \frac{\delta}{s}\right)\frac{\Delta K}{K}. \tag{5.9}$$

For both time-dependent and time-independent sensitivity changes ΔK, it is obvious that the sensor output signal is directly influenced by these deviations. In case of temporal drift effects, the relative deviation $\varepsilon_{\mathrm{d}}^{\mathrm{defl}}(s)$ of the sensor output is time-dependent. For infinite measurement times, the deviation reaches infinity. On the other hand, Eq. (5.9) becomes Eq. (5.6) for $\delta = 0$.

5.2.3 Response time

Eq. (5.7) describes the time-dependence of the quasi-static sensor parameter K and, hence, characterizes the unwanted long-term behavior of changing sensor properties, whereas $G(s)$ describes the dynamic short-term behavior. Usually, a sensor response time as short as possible is desired. As quoted in Table 5.2, sensors can be classified by their transfer function $G(s)$. In general, the dynamic behavior of a sensor depends on the location of the roots of the numerator polynomial (zeros) and the denominator polynomial (poles) in the complex GAUSS-plane. In particular, the sensor response time is determined by the dominant pole that is placed closest to the imaginary axis. The poles and zeros are preset by the sensor coefficients (a_0, a_1, a_2, b_0, b_1) and consequently by the intrinsic sensor properties. Therefore, the operation of a sensor in deflection mode (open-loop configuration) leads to a defined dynamic response that cannot be modified.

5.3 General sensor model for compensation method

5.3.1 System model

Sensors using the compensation method exhibit a closed-loop feedback structure (Fig. 5.6) where the deflection is brought back to zero by a compensation force. This causes a force equilibrium and keeps the deflection of the sensor – occasionally with a small remnant control deviation – at zero. The deflection cannot longer be the measurement signal because it amounts more or less to zero. The sensor signal being proportional to the measurand is now the compensation force or an electrical quantity creating this compensation force. The resulting compensation circuit must at least contain the sensor itself, an amplifier, and the controller/actuator unit. Other components (e. g. D/A converter, filter) may be necessary, but do not alter the general system behavior and are thus neglected in the following considerations. Because we act on the assumption of an electrical sensor output signal, amplification with a high-gain amplifier is assumed. For the sake of simplicity, we assume the amplifier, actuator, and controller to be much faster than the sensor itself. This would be particularly the case for e. g. mechanical and chemical sensors. Hence, we can consider the actuator and the controller as one block placed in the feedback path. For other sensor types, where this simplifying assumption does not hold, the closed-loop sensor model with only one block in the feedback path is nevertheless applicable. In this case, the dominant time constant of the feedback block is a combination of the time constants of the actuator and the controller parameters, respectively.

Note that the input signal X_i' of the feedback structure is not necessarily the measuring quantity X_i. If a direct compensation of X_i is not possible, because the desired quantity cannot be generated by an existing device or the generation is too complicated, an equivalent quantity X_i', which is correlated to X_i as $X_i' = F \cdot X_i$ and which is accessible for compensation must exist. This is a necessary requirement for the applicability of the compensation method to a certain sensor. The correlation

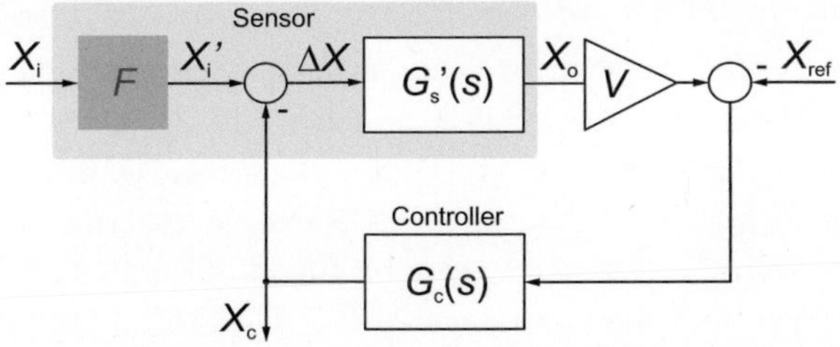

Figure 5.6: Closed-loop compensation circuit where the sensor output signal X_o is fed back to the sensor input via an amplifier and a controller. F is a transfer element transforming an arbitrary input quantity in a quantity which can be actually compensated.

between X_i and X_i' can be a simple factor (e. g. the gravity of Earth connecting mass and force in a balance) or rather complex (e. g. broadband lambda-probe or closed-loop hydrogel-based sensors (Chapter 6)) depending on the specific sensor. In any case, the overall system behavior of the compensation circuit is independent of F. Therefore, to keep simplicity without loosing generality, we assume $F = 1$ and, thus, $X_\mathrm{i}' = X_\mathrm{i}$ and $G_\mathrm{s}' = G_\mathrm{s}$ for all further considerations.
Initially, the input signal X_i causes the sensor output signal X_o. X_o is amplified to $V \cdot X_\mathrm{o}$, the preset reference value X_ref is subtracted from it and fed back to the controller that generates the compensation quantity X_c. Hence, X_c counteracts the sensor input signal. Assuming ideal compensation, the sensor input $X_\mathrm{i} - X_\mathrm{c}$ is balanced during the whole measuring process in such a way that ΔX and, hence, $V \cdot X_\mathrm{o} - X_\mathrm{ref}$ become zero at the steady state. Hence, the sensor output X_o remains constant independent of changes of the input signal. However, the "force" X_c, which is necessary to compensate changes of the input signal, depends on the input signal. Therefore, X_c is a measure of the input signal X_i and, thus, the output signal of the "force"-compensated sensor. The resulting transfer function of the compensation circuit can then be written as

$$G_\mathrm{comp}(s) = \frac{X_\mathrm{c}(s)}{X_\mathrm{i}(s)} = \frac{V \cdot G_\mathrm{s}(s) \cdot G_\mathrm{c}(s)}{1 + V \cdot G_\mathrm{s}(s) \cdot G_\mathrm{c}(s)} = \frac{\tilde{G}(s)}{1 + \tilde{G}(s)}, \tag{5.10}$$

with

$$\tilde{G}(s) = V \cdot G_\mathrm{s}(s) \cdot G_\mathrm{c}(s) \tag{5.11}$$

and expressed in a more general form as

$$G_\mathrm{comp}(s) = \frac{N_\mathrm{comp}(s)}{D_\mathrm{comp}(s)} = \frac{c_m s^m + ... + c_1 s + c_0}{d_n s^n + ... + d_1 s + d_0}, \tag{5.12}$$

where both numerator N_comp and denominator D_comp are composed of polynomials. From Eq. (5.10) one can see that for $G_\mathrm{comp} \neq 0$, the compensation quantity X_c can be used as a direct measure of the measuring quantity X_i. Because of the feedback loop, a transfer function in the form $\tilde{G}/(1 + \tilde{G})$ arises (cp. Table 5.1) which exibits a completely different structure compared to the simple open-loop transfer function. Moreover, the system properties can now be tuned and adjusted in certain limits via the amplifier gain and the parameters of G_c. Thus, G_comp can be regarded as a variably tunable transfer function.

The controller is assumed to be a standard linear P-, PI-, or PID- controller, respectively. In principle, a specific controller could be designed for the task of

Table 5.3: Time-dependent characteristics of the controller output signal $x_{\mathrm{c}}(t)$ as a function of the controller input signal $e_{\mathrm{c}}(t)$ and the corresponding transfer functions of basic linear standard controller types [188].

Controller type	Control law	Transfer function
P-controller	$x_{\mathrm{c}}(t) = k_{\mathrm{P}} e_{\mathrm{c}}(t)$	$G_{\mathrm{c}}^{\mathrm{P}}(s) = k_{\mathrm{P}}$
PI-controller	$x_{\mathrm{c}}(t) = k_{\mathrm{P}} e_{\mathrm{c}}(t) + k_{\mathrm{I}} \int_0^t e_{\mathrm{c}}(t')\mathrm{d}t'$	$G_{\mathrm{c}}^{\mathrm{PI}}(s) = k_{\mathrm{P}} + \frac{k_{\mathrm{I}}}{s}$
PID-controller	$x_{\mathrm{c}}(t) = k_{\mathrm{P}} e_{\mathrm{c}}(t) + k_{\mathrm{I}} \int_0^t e_{\mathrm{c}}(t')\mathrm{d}t' + k_{\mathrm{D}} \frac{\mathrm{d}e_{\mathrm{c}}(t)}{\mathrm{d}t}$	$G_{\mathrm{c}}^{\mathrm{PID}}(s) = k_{\mathrm{P}} + \frac{k_{\mathrm{I}}}{s} + k_{\mathrm{D}} \cdot s$

compensation. However, practically, the predominant majority of all control tasks are realized by standard controllers [189]. Therefore, we focus on those standard controllers. The transfer functions of the three controller types are listed in Table 5.3. Here, k_{P} is the P-factor, k_{I} the I-factor, and k_{D} the D-factor with

$$k_{\mathrm{I}} = \frac{k_{\mathrm{P}}}{T_{\mathrm{I}}} \tag{5.13}$$

and

$$k_{\mathrm{D}} = k_{\mathrm{P}} T_{\mathrm{D}} \tag{5.14}$$

where T_{I} and T_{D} are the integral time and the differential time of the controller, respectively. We assume k_{P}, k_{I}, and k_{D} to be constant. The application of the controller transfer functions $G_{\mathrm{c}}(s)$ from Table 5.3 and the sensor transfer functions $G_{\mathrm{s}}(s)$ from Table 5.2 to Eq. (5.10) leads to the closed-loop transfer functions of the three sensor types. The resulting transfer functions of the different sensor-controller combinations are listed in Table 5.4 where $K^* = K \cdot V$. In the following, $K^* \cdot k_{\mathrm{P}}$ is termed as feedback factor. However, the most important requirement for a technical use of a closed-loop system is the system stability, i. e. a bounded input signal leads to a bounded output signal (BIBO-stability) [184]. Stability has to be guaranteed for all operating conditions. This necessitates the analysis of the closed-loop transfer functions to determine the parameter range in which the system is stable.

Table 5.4: Closed-loop transfer functions $G_{\mathrm{comp}}(s)$ for all sensor-controller combinations considered; $K^* = K \cdot V$.

$G_{\mathrm{comp}}(s)$	P	PI	PID
I: PT_1	$\frac{K^* k_\mathrm{P}}{sT + (K^* k_\mathrm{P} + 1)}$	$\frac{K^*(k_\mathrm{P} s + k_\mathrm{I})}{s^2 T + (1 + K^* k_\mathrm{P}) s + K^* k_\mathrm{I}}$	$\frac{K^*(k_\mathrm{D} s^2 + k_\mathrm{P} s + k_\mathrm{I})}{(T + K^* k_\mathrm{D}) s^2 + (1 + K^* k_\mathrm{P}) s + K^* k_\mathrm{I}}$
II: PT_2	$\frac{K^* k_\mathrm{P}}{T_1^2 s^2 + T_2 s + (1 + K^* k_\mathrm{P})}$	$\frac{K^*(k_\mathrm{P} s + k_\mathrm{I})}{T_1^2 s^3 + T_2 s^2 + (1 + K^* k_\mathrm{P}) s + K^* k_\mathrm{I}}$	$\frac{K^*(k_\mathrm{D} s^2 + k_\mathrm{P} s + k_\mathrm{I})}{T_1^2 s^3 + (T_2 + K^* k_\mathrm{D}) s^2 + (1 + K^* k_\mathrm{P}) s + K^* k_\mathrm{I}}$
III: PT_2-D	$\frac{K^* k_\mathrm{P} T_3 s}{T_1^2 s^2 + (T_2 + K^* k_\mathrm{P} T_3) s + 1}$	$\frac{K^* T_3(k_\mathrm{P} s + k_\mathrm{I})}{T_1^2 s^2 + (T_2 + K^* k_\mathrm{P} T_3) s + (1 + K^* k_\mathrm{I} T_3)}$	$\frac{K^* T_3(k_\mathrm{D} s^2 + k_\mathrm{P} s + k_\mathrm{I})}{(T_1^2 + K^* k_\mathrm{D} T_3) s^2 + (T_2 + K^* k_\mathrm{P} T_3) s + (1 + K^* k_\mathrm{I} T_3)}$

5.3.2 Stability requirements

Stability analysis is easily accessible in the frequency domain by analysis of the closed-loop transfer function $G_{\mathrm{comp}}(s)$ as stated in Eq. (5.12). This transfer function is asymptotically stable if and only

- if G_{comp} is proper, i. e. the degree m of the numerator N_{comp} is smaller or equal to the degree n of the denominator D_{comp} ($m \leq n$, see Eq. (5.12)), and
- if all poles of G_{comp} are placed in the complex left open half-plane (LOH).

The first stability condition can be easily evaluated by a simple look at the respective closed-loop transfer function (Table 5.4). The second stability condition can be verified by a stability criterion (e. g. ROUTH criterion [184] as used within this work) applied to the denominator polynomial $\mathrm{D}_{\mathrm{comp}}(s)$. Here, the necessary condition is that all coefficients d_ν of the denominator have the same sign (STODOLA condition [190, 191]) (e. g. $d_m, \ldots, d_0 > 0$). The sufficient condition results from the employed stability criterion and has to be evaluated subsequently. If the numerator degree amounts to $n \leq 2$, the STODOLA condition is also sufficient. For that simplified case, no stability criterion has to be used in addition.

Beside the fundamental necessity for system stability, we additionally claim a minimum-phase system. Thus, all zeros of G_{comp} have to be located in the complex left half-plane, including the imaginary axis. In this case, the same criterion (second criterion above) as for the poles is valid, except that the coefficients c_ν of the numerator can be equal to zero as well ($c_m, \ldots, c_0 \geq 0$). A closed-loop transfer function exhibiting nonminimum-phase zeros (zeros in the open right half plane) will show initial overshoot or undershoot, i. e. the step response initially shows an inverse response, spending part of its time going in the "wrong" direction until it changes its direction towards the steady state. This kind of step response exhibits non-monotonic behavior. This inverse response exists, if the transfer function has at least one zero in the right half-plane. Depending on the number of zeros in the right half-plane, the inverse response can become quite complex and can exhibit more direction reversals and zero-crossings (time-dependent signal passes through the value of zero) [192, 193]. This behavior is inadequate for sensor applications because the steady state should be reached as fast as possible and a monotone step response is aspired. Therefore, the controller parameters have to be chosen in such a manner that the asymptotically stable closed-loop transfer function shows minimum-phase behavior at the same time. From Table 5.4 it is apparent that the first stability criterion is fulfilled for all transfer functions. For real sensors, $K > 0$, $V > 0$, and $T > 0$ can be assumed without limitation. Using these assumptions, first the range of the control parameters which guarantee BIBO-stability is calculated by

Table 5.5: Parameter range guaranteeing BIBO-stability and a minimum-phase system as well as the resulting combined parameter range satisfying both requirements for all considered sensor-controller combinations.

$G_{\text{comp}}(s)$	P	PI	PID
		Parameter range for a BIBO-stable system	
I: PT_1	$k_P > -\frac{1}{K^*}$	$k_P > -\frac{1}{K^*}$, $k_I > 0$	$k_P > -\frac{1}{K^*}$, $k_I > 0$, $k_D > -\frac{T}{K^*}$
II: PT_2	$k_P > -\frac{1}{K^*}$	$k_P > -\frac{1}{K^*}$, $k_I > 0$	$k_P > -\frac{1}{K^*}$, $k_I > 0$, $k_D > -\frac{T_2}{K^*}$
		$1 + K^* k_P - \frac{T_1^2 K^* k_I}{T_2} > 0$ [a]	$1 + K^* k_P - \frac{T_1^2 K^* k_I}{T_2 + K^* k_D} > 0$ [a]
III: PT_2-D	$k_P > -\frac{T_2}{K^* T_3}$	$k_P > -\frac{T_2}{K^* T_3}$, $k_I > -\frac{1}{K^* T_3}$	$k_P > -\frac{T_2}{K^* T_3}$, $k_I > -\frac{1}{K^* T_3}$, $k_D > -\frac{T_1^2}{K^* T_3}$
		Parameter range for a minimum-phase system	
I-III:	$k_P \geq 0$	$k_P \geq 0, k_I \geq 0$	$k_P \geq 0$ [b], $k_I \geq 0, k_D \geq 0$
		Resulting combined parameter range	
I: PT_1	$k_P > 0$ [c]	$k_P \geq 0$, $k_I > 0$	$k_P > 0$ [d], $k_I > 0$, $k_D \geq 0$
II: PT_2	$k_P > 0$	$k_P \geq 0$, $k_I > 0$	$k_P > 0$, $k_I > 0$, $k_D \geq 0$
		$T_2(1 + K^* k_P) > T_1^2 K^* k_I$ [a]	$(T_2 + K^* k_D)(1 + K^* k_P) > T_1^2 K^* k_I$ [a]
III: PT_2-D	$k_P > 0$	$k_P \geq 0$, $k_I \geq 0$	$k_P > 0$ [d], $k_I \geq 0$, $k_D \geq 0$

[a]This criterion results from the application of the Routh criterion.
[b]Except for $k_D = 0$, this case is however not relevant due to the stability requirement.
[c]The case $k_P = 0$ is not relevant for sensor applications.
[d]$k_P > 0$ instead of $k_P \geq 0$ excludes the imaginary zero pair.

utilization of the second stability criterion. Subsequently, the parameter range to assure a minimum-phase behavior is calculated. These results as well as the resulting parameter range meeting both requirements are listed in Table 5.5. As can be seen in Table 5.5, for the combination of a sensor with PT_2 behavior together with a PI- or PID-controller, respectively, the degree of the denominator is of order three. Because of that, the Routh criterion derived from the Routh table has been used to get the sufficient stability condition. Stability analysis by utilization of the Routh criterion is explained in more detail in Appendix B.1. From these considerations it comes out that the range of controller parameters guaranteeing stability depends on the sensor properties. Note that the second stability condition is even satisfied if certain parameters exhibit negative values. However, if one claims stability and minimum-phase behavior at the same time, the controller parameters do not depend on the sensor properties. The parameters can be any positive value or even equal zero, except for the combination of a PT_2 element with a PI- or a PID-controller. This enables a wide range of possible controller parameters for the adjustment of the desired system behavior for specific applications. Moreover, these findings do not show any contradiction towards the application of the compensation method in a technical measurement system. However, from a practical point of view, the static feedback gain cannot be increased ad infinitum. Above a certain critical gain, a finite lag of the feedback path will lead to instabilities.

5.3.3 Steady-state behavior

a. Static deviation from ideal compensation

The designated aim of the compensation method is an ideal compensation of the sensor input X_i such that $V \cdot X_\mathrm{o} - X_\mathrm{ref}$ becomes zero at the steady state ($s \to 0$, $t \to \infty$). For the sake of simplicity, let us assume $X_\mathrm{ref} = 0$. Then,

$$\Delta X = X_\mathrm{i} - X_\mathrm{c} = X_\mathrm{i} - (V \cdot G_\mathrm{s} \cdot G_\mathrm{c} \cdot \Delta X) \tag{5.15}$$

should be consequently as small as possible. From Eq. (5.15) we can set up a transfer function G^Δ between the compensation deviation ΔX and the input signal X_i as

$$G^\Delta := G_{X_\mathrm{i}}^{\Delta X} = \frac{\Delta X(s)}{X_\mathrm{i}(s)} = \frac{1}{1 + V \cdot G_\mathrm{s} \cdot G_\mathrm{c}}. \tag{5.16}$$

Since Eq. (5.16) comprises the same denominator as the general transfer function in Eq. (5.10), we can act on the assumption that all poles of G^Δ exhibit negative real part (for the parameter range in Table 5.5) and that the transfer function is thus stable. In this case, one can apply the final value theorem of the LAPLACE transform, generally written as

$$\lim_{t \to \infty} x(t) = \lim_{s \to 0} [s \cdot X(s)]. \tag{5.17}$$

Using Eq. (5.17), we can calculate the deviation $\Delta x(t \to \infty)$ from ideal compensation in the steady state for a step change of X_i as

$$\Delta x(\infty) = \lim_{t \to \infty} \Delta x(t) = \lim_{s \to 0} \left[s \cdot G^\Delta(s) \cdot \frac{1}{s} \right] = \lim_{s \to 0} G^\Delta(s). \tag{5.18}$$

Example: PT_1-type sensor with P-controller

For the simplest example of a PT_1-type sensor in combination with a P-controller, G^{Δ} is

$$G^{\Delta}(s) = \frac{sT + 1}{sT + K^{*}k_{\mathrm{P}} + 1}, \tag{5.19}$$

leading to

$$\Delta x(\infty) = \lim_{s \to 0} \left(\frac{sT + 1}{sT + K^{*}k_{\mathrm{P}} + 1} \right) = \frac{1}{1 + K^{*}k_{\mathrm{P}}}. \tag{5.20}$$

The static deviation from ideal compensation for all sensor-controller combinations are given in Table 5.6. Here, one can see the same behavior for both PT_1- and PT_2-type sensors regarding the static deviation from ideal compensation. For application of a P-controller, Δx depends on the feedback factor. For unity feedback ($K^{*}k_{\mathrm{P}} = 1$), a deviation of 50 % is observed. However, for an increasing feedback factor, the deviation decreases and converges to zero for a high-gain feedback. A deviation of only 1 % is already reached with a feedback factor of 100. For the application of a PI- or PID-controller, Δx turns out to be zero and, thus, independent of the feedback parameters. Therefore, a PI- or PID-controller should be used for these sensor types in order to achieve full compensation. The combination of a PT_2-D-type sensor with a P-controller cannot be used for sensor applications. This can be easily understood by application of the final value theorem to the respective transfer function in Table 5.4. In this case, the sensor output at the steady state always equals zero independent of the input signal. Thus, different input signals would always lead to the same (zero) output signal. However, a PI- or a PID-controller can be successfully applied. The application of PI- or PID-controller to this sensor type leads to a static deviation which depends on the feedback factor, the integral time T_{I}, and the time constant T_2. Thus, for a high feedback factor and a small integral time, the static deviation converges to zero.

Table 5.6: Static deviation from ideal compensation for all considered sensor-controller combinations.

$\Delta x(\infty)$	P	PI	PID
PT_1	$\frac{1}{1 + K^{*}k_{\mathrm{P}}}$	0	0
PT_2	$\frac{1}{1 + K^{*}k_{\mathrm{P}}}$	0	0
PT_2-D	N.A.	$\frac{1}{K^{*}T_2k_{\mathrm{I}} + 1}$	$\frac{1}{K^{*}T_2k_{\mathrm{I}} + 1}$

b. Influence of static parameter variations

For a static analysis of the compensation circuit, we assume the closed-loop system to be in a steady state ($G_{\text{comp}}^{\text{stat}} = G_{\text{comp}}(s \to 0)$). So far, we have considered an ideal sensitivity K for the transfer function of the compensation circuit. However, as described in the general sensor model in Section 5.2.1, the sensitivity is usually influenced by the interference quantity. Furthermore, compared to the open-loop sensor, the amplifier and the controller parameters can exhibit deviations from ideal behavior as well. Considering small static variations, the impact of these deviations on the output signal X_{c} can be calculated by a first-order TAYLOR-series as

$$\varepsilon_{\text{s}}^{\text{comp}} = \frac{\Delta X_{\text{c}}}{X_{\text{c}}} = \frac{1}{X_{\text{c}}}\left(\frac{\partial X_{\text{c}}}{\partial K} \cdot \Delta K + \frac{\partial X_{\text{c}}}{\partial V} \cdot \Delta V + \sum_{i=1}^{3} \frac{\partial X_{\text{c}}}{\partial k_i} \cdot \Delta k_i\right). \tag{5.21}$$

Here, $k_i \in \{k_{\text{P}}, k_{\text{I}}, k_{\text{D}}\}$ and $\Delta k_i \in \{\Delta k_{\text{P}}, \Delta k_{\text{I}}, \Delta k_{\text{D}}\}$ are the controller parameters and their systematic deviations, respectively, depending on which controller type is used.

Example: PT_1-type sensor with P-controller

For the simplest example of a PT_1-type sensor in combination with a P-controller, $\varepsilon_{\text{s}}^{\text{comp}}$ is

$$\begin{aligned}\varepsilon_{\text{s}}^{\text{comp}} &= \frac{1 + KVk_{\text{P}}}{KVk_{\text{P}}}\left[\frac{Vk_{\text{P}}}{(1+KVk_{\text{P}})^2}\Delta K + \frac{Kk_{\text{P}}}{(1+KVk_{\text{P}})^2}\Delta V + \frac{VK}{(1+KVk_{\text{P}})^2}\Delta k_{\text{P}}\right] \\ &= \frac{1}{1+K^* k_{\text{P}}}\left(\frac{\Delta K}{K} + \frac{\Delta V}{V} + \frac{\Delta k_{\text{P}}}{k_{\text{P}}}\right)\end{aligned} \tag{5.22}$$

To allow an easy comparison with the open-loop sensor, the deviations which do not result from the sensor itself can be summarized as a feedback deviation ΔFB and can generally be described as a multiple (β) of the sensor deviation as

$$\frac{\Delta FB}{FB} = \frac{\Delta V}{V} + \frac{\Delta k_{\text{P}}}{k_{\text{P}}} = \beta \cdot \frac{\Delta K}{K}. \tag{5.23}$$

Thus, Eq. (5.22) becomes

$$\varepsilon_{\text{s}}^{\text{comp}} = \frac{1+\beta}{1+K^* k_{\text{P}}}\frac{\Delta K}{K}. \tag{5.24}$$

The ratio between the relative systematic static deviation of the closed-loop compensation circuit and the open-loop sensor is

$$\frac{\varepsilon_{\mathrm{s}}^{\mathrm{comp}}}{\varepsilon_{\mathrm{s}}^{\mathrm{defl}}} = \frac{1+\beta}{1+K^{*}k_{\mathrm{P}}}. \tag{5.25}$$

This ratio is illustrated in Fig. 5.7 as a function of the feedback factor and β indicating the deviation contribution of the feedback elements.

Usually, the feedback elements should be designed such that systematic deviations are much smaller than the deviations of the sensor itself. As can be seen from Fig. 5.7, if the feedback deviations are just 1 % ($\beta = 0.01$) of the sensor deviations, a unity feedback already minimizes the overall deviation to 50 % compared to an open-loop sensor. However, if the circuit comprises higher feedback deviations, the feedback factor can be systematically increased to achieve a specified lower deviation of the output signal. If one wants to achieve a constant uncertainty, the required feedback factor linearly increases with β as stated in Eq. (5.25).

The ratio between the deviations of the closed-loop configuration compared to the open-loop sensor for all sensor-controller combinations are listed in Table 5.7. From this table, one can see that for PT_1- and PT_2-type sensors in combination with a P-controller, the systematic deviations depend on the deviation contributions of the single circuit elements and the feedback factor. However, by additional application of an I-part (PI- or PID-controller), the systematic deviations can be completely eliminated and do not depend on the deviations of the circuit elements any longer. For a PT_2-D-sensor in combination with a PI- or PID-controller, a similar behavior as discussed for the P-controller before can be observed. Here, the overall deviation of the output signal cannot only be tuned by the static amplification V and the

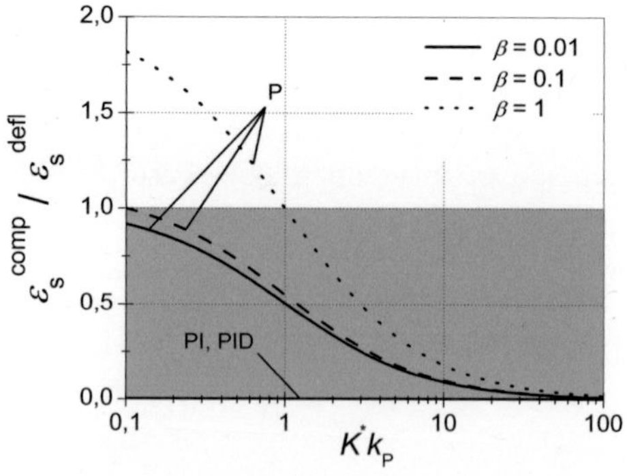

Figure 5.7: Ratio between systematic static deviations of the closed-loop sensor and the open-loop sensor versus the feedback factor $K^{*}k_{\mathrm{P}}$ for a PT_1-type sensor in combination with all three controller types. β indicates different deviation contributions from the feedback elements compared to the deviation of the sensor itself.

Table 5.7: Systematic static deviations of the closed-loop sensor for all sensor-controller combinations.

ε_s^{comp}	P	PI	PID
PT_1	$\frac{\frac{\Delta K}{K}+\frac{\Delta V}{V}+\frac{\Delta k_P}{k_P}}{1+K^*k_P}$	0	0
PT_2	$\frac{\frac{\Delta K}{K}+\frac{\Delta V}{V}+\frac{\Delta k_P}{k_P}}{1+K^*k_P}$	0	0
PT_2-D	N.A.	$\frac{\frac{\Delta K}{K}+\frac{\Delta V}{V}+\frac{\Delta k_I}{k_I}}{1+K^*k_P\frac{T_3}{T_I}}$	$\frac{\frac{\Delta K}{K}+\frac{\Delta V}{V}+\frac{\Delta k_I}{k_I}}{1+K^*k_P\frac{T_3}{T_I}}$

P-factor k_P, but additionally by the integral time T_I. The smaller the integral time is, the better is the resulting suppression of deviations.

c. Influence of time-dependent parameter variations

If the external interference quantity varies with time, the long-term sensor output signal shows a continuous drift even if the dynamic sensor part is already equilibrated. The dynamics of the drift process is modelled by the transfer function $\tilde{K}(s)$. The application of the drift model from Eq. (5.7) to the closed-loop compensation circuit is illustrated in Fig. 5.8.

The closed-loop transfer functions according to Eq. (5.10), considering all three controller types, are calculated and listed in Table 5.8.

From Table 5.8 one can observe that the unstable transfer function of the open-loop sensor exhibiting drift in Eq. (5.7) is transformed into a more complex transfer function. Here, the coefficients of the numerator and the denominator can be tuned. We also find that all three closed-loop transfer functions are exactly proper. The parameter range enabling a stable as well as a minimum-phase system is determined according to the criteria in section 5.3.2 and listed in Table B.1. From this table one can see that the system can be operated in a broad parameter range without

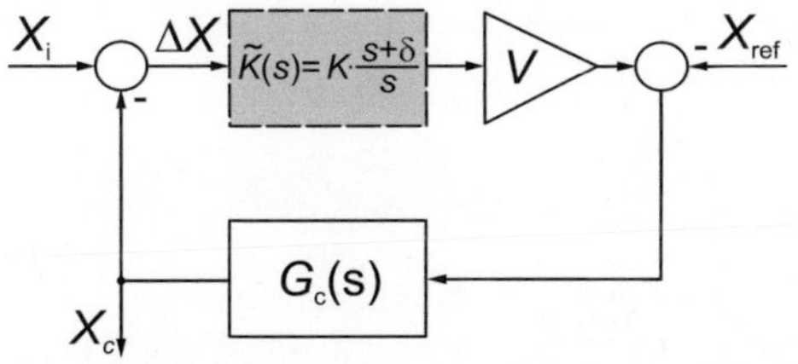

Figure 5.8: Closed-loop compensation circuit where the quasi-static sensor part is characterized by a sensitivity $\tilde{K}$, describing the dynamics of the drift process by a drift model.

Table 5.8: Closed-loop transfer function $G_{\text{comp}}^{\Delta\text{K,d}}$ considering a time-dependent sensitivity $\tilde{K}$ instead of an ideal static sensitivity K by application of the drift model from Eq. (5.7) to the static sensor part (gray box in Fig. 5.8).

	$G_{\text{comp}}^{\Delta\text{K,d}}(s)$
P	$\dfrac{K^* k_\text{P}(s+\delta)}{(1+K^* k_\text{P})s + K^* k_\text{P}\delta}$
PI	$\dfrac{K^* k_\text{P} s^2 + (K^* k_\text{I} + K^* k_\text{P}\delta)s + K^* k_\text{I}\delta}{(1+K^* k_\text{P})s^2 + (K^* k_\text{I} + K^* k_\text{P}\delta)s + K^* k_\text{I}\delta}$
PID	$\dfrac{K^* k_\text{D} s^3 + (K^* k_\text{P} + K^* k_\text{D}\delta)s^2 + (K^* k_\text{I} + K^* k_\text{P}\delta)s + K^* k_\text{I}\delta}{K^* k_\text{D} s^3 + (1 + K^* k_\text{P} + K^* k_\text{D}\delta)s^2 + (K^* k_\text{I} + K^* k_\text{P}\delta)s + K^* k_\text{I}\delta}$

jeopardizing the system stability. For the utilization of a P- or PI-controller, the controller parameters have to be simply positive to assure a stable minimum-phase system. For a PID-controller, the restrictions seem more complicated. However, if one simplifies the expression arising from the ROUTH table, one finds that again all of the stated requirements are met for positive controller parameters (k_P, k_I, $k_\text{D} > 0$) independent of the drift-rate (see Section B.2.1 in Appendix B.2). In this sense, stability directly involves a system without drift, because a system exhibiting drift is *a priori* unstable. This can be shown by calculating the steady state behavior of the output signal (x_c) considering a unit step input according to Eq. (5.17) as

$$x_\text{c}(\infty) = \lim_{t\to\infty} x_\text{c}(t) = \lim_{s\to 0} G_{\text{comp}}^{\Delta\text{K,d}}(s) = 1 \tag{5.26}$$

and is valid for all three controller types. One can see that the output signal of the closed-loop system is constant at the steady state for all three controller types, whereas the output signal of the open-loop sensor diverges to infinity (cp. Fig. 5.5).

To enable a quantitative comparison of the quality of drift suppression for the three different controller types in dependence of the controller parameters, the use of a global performance index is reasonable. A performance index, which is easy to handle and meaningful at the same time, is the Integral of Squared Error (ISE) [194] which can be written in the time domain as

$$J = \int_0^\infty \Delta x^2(t)\text{d}t. \tag{5.27}$$

Here, $\Delta x(t)$ is the difference between the input step $x_\text{i} = 1(t)$ and the actual sensor output signal x_c (cp. Fig. 5.8). This integral is a function of the controller parameters ($J = f(k_\text{P}, k_\text{I}, k_\text{D})$). The requirement for best drift suppression is that J is as small as possible ($J \overset{!}{=} \min$). For a quantitative evaluation of this requirement and to get reasonable analytical equations for J, the transformation from the time to the frequency domain is reasonable. Since we already proved system stability for all three controller types, we can use PARSEVAL's theorem [194] for the accomplishment of the transformation as

$$J = \int_0^\infty \Delta x^2(t)\mathrm{d}t = \frac{1}{2\pi}\int_{-\infty}^{+\infty} |\Delta X(s)|^2 \mathrm{d}s = \frac{1}{2\pi}\int_{-\infty}^{+\infty} \Delta X(s)\Delta X(-s)\mathrm{d}s \tag{5.28}$$

with

$$\Delta X(s) = \frac{1}{1 + K\frac{s+\delta}{s}VG_\mathrm{c}} \cdot \frac{1}{s} = \frac{k_0 + k_1 s + ... + k_{r-1}s^{r-1}}{l_0 + l_1 s + ... + l_r s^r},\ l_r \neq 0 \tag{5.29}$$

according to the system structure in Fig. 5.8 for a unit step input. The explicit general solution of this integral up to $r = 10$ can be found in [194]. The insertion of the respective controller transfer function into Eq. (5.29) gives the compensation deviation $\Delta X(s)$ as stated in Table B.2. The solution of the integral according to Eq. (5.28) and [194], respectively, leads to the quality criteria J for the three controller types as given in Table 5.9 (cp. Section B.2.2). From Tab. 5.9 it is obvious that the equation

$$\frac{\partial J}{\partial k_i} = 0, \text{ with } k_i \in \{k_\mathrm{P}, k_\mathrm{I}, k_\mathrm{D}\} \tag{5.30}$$

cannot be explicitly satisfied. No local minimum in dependence on the controller parameters can be found because the first derivatives ($\partial J/\partial k_\mathrm{P}$, $\partial J/\partial k_\mathrm{I}$, $\partial J/\partial k_\mathrm{D}$) only converge to zero for infinitely large controller parameters (k_P, k_I, $k_\mathrm{D} \to \infty$). This is not in contradiction to the stability requirements and means that the drift suppression capability of the respective controller type is only restricted by practical limitations of the controller parameters. To exemplarily illustrate the drift suppression of the closed-loop sensor for different controller types, the sensor step responses for selected controller parameters as well as the performance index as a function of the respective controller parameters are illustrated in Fig. 5.9.

From Fig. 5.9 (left) it can be observed that, when the system experiences an input step, it instantaneously follows this step to a certain extent until it is damped and converges to unity. The higher the respective controller parameters are, the less damping can be observed. Furthermore, the steady state is reached even earlier. Here, reaching the steady state means that the sensor signal sets in into a tolerance band of $\pm 1\,\%$ of $x_\mathrm{c}(t) = 1$ (gray region) and does not leave it anymore. This trend also reflects in the ISE performance indices in Fig. 5.9 (right). Here, generally, J is

Table 5.9: Performance index from Eq. (5.28) for drift suppression for the three controller types.

	Performance index J
P	$\frac{1}{2K^*k_\mathrm{P}\delta(1+K^*k_\mathrm{P})}$
PI	$\frac{1}{2K^*k_\mathrm{P}\delta(1+K^*k_\mathrm{P})+2K^*k_\mathrm{I}(1+K^*k_\mathrm{P})}$
PID	$\frac{1}{2K^*k_\mathrm{P}\delta(1+K^*k_\mathrm{P})+2K^*k_\mathrm{I}(1+K^*k_\mathrm{P})+2K^{*2}k_\mathrm{P}k_\mathrm{D}\delta^2}$

minimized for increasing controller parameters. Moreover, with increasing controller complexity, J is additionally minimized. However, it can be seen that the P-part has the strongest influence on the quality of drift suppression. It is noteworthy that even a feedback factor of $K^*k_\mathrm{P}=1$ leads to a complete drift-suppression solely due to the closed-loop sensor structure. However, tuning of the feedback factor leads to a faster accomplishment of the steady state. The I-part shows less influence on the overall drift suppression but can nevertheless be used to improve the time to reach equilibrium. For small k_I an overshoot can be observed due to the zeros of the numerator in the transfer function (Fig. 5.9b (left)). However, for increasing k_I, this overshoot decreases. For a PID-controller (Fig. 5.9c (right)), the influence of the I-part is more pronounced than that of the D-part. Both parts show higher influences for small feedback factors. However, for increasing feedback factors, the influence of these parts decreases and becomes negligible for very high feedback factors. The I- and the D-part however influence the signal characteristics. The D-part even leads to decaying oscillations within the tolerance band (Fig. 5.9c (left)) which is not an aspired behavior. Thus, we can conclude that the use of a P-controller with high feedback gain can satisfactorily fulfill the demand of drift suppression. A PI-controller is recommendable, if a faster accomplishment of the steady state is aspired, or if the feedback factor cannot be increased due to technical limitations. However, the use of a PID-controller is basically possible but would not be very convenient.

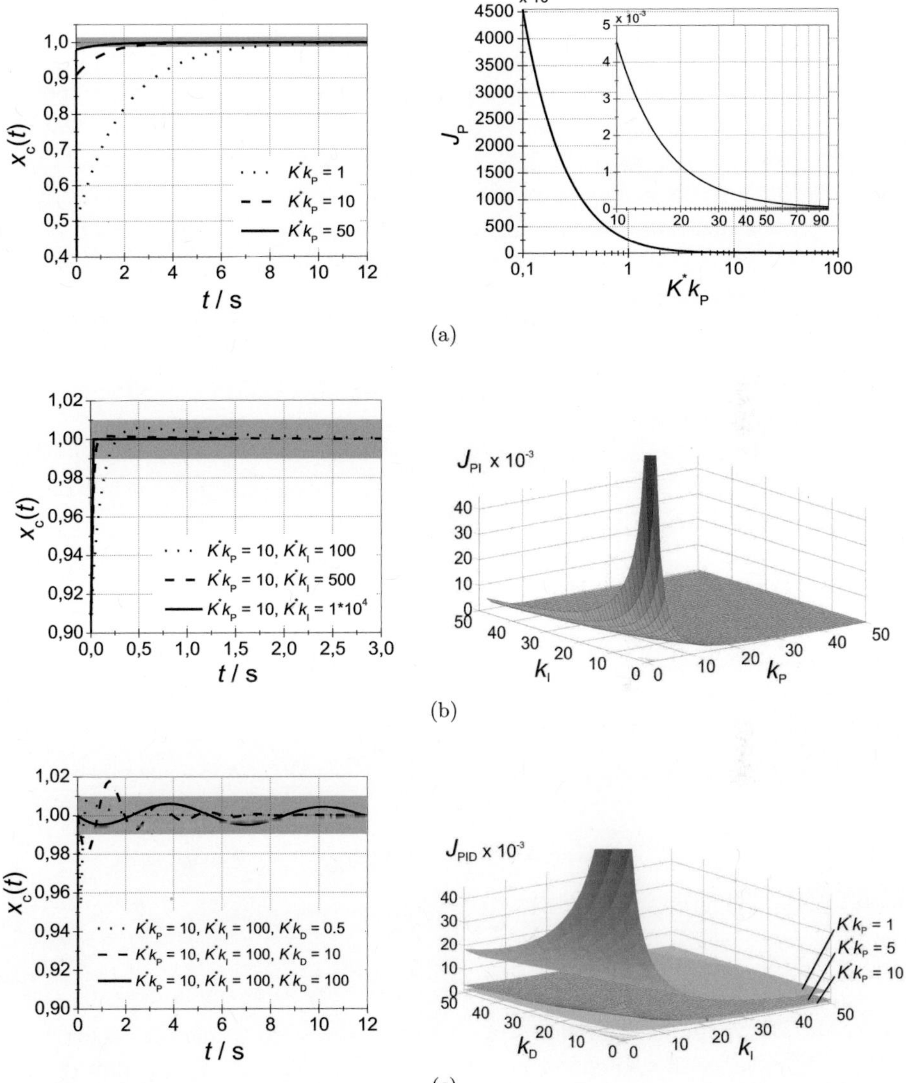

Figure 5.9: Step response of the closed-loop sensor for selected controller parameters (left) and ISE performance index as a function of the respective controller parameter (all other parameters are fixed) (right) for a (a) P-controller, (b) PI-controller, and (c) PID-controller. The light gray region indicates a tolerance band of $\pm 1\,\%$ of the steady state value. For all graphs a drift rate of $\delta = 1\,s^{-1}$ was considered.

5.3.4 Dynamic system behavior

a. Response time

In general, the sensor response time of any sensor type is determined by the poles of its transfer function. In the simplest case, only one pole and, thus, only one time constant, exists. However, for higher-order transfer functions, several poles do occur. The pole that is located nearest to the imaginary axis in the LOH is dominant and determines the time to reach equilibrium. The condition for response time reduction is that the dominant pole of the closed-loop system is located in the LOH relative to the dominant pole of the open-loop sensor

$$\Re\{s_{\mathrm{p}}^{\mathrm{comp}}\} < \Re\{s_{\mathrm{p}}^{\mathrm{defl}}\} \tag{5.31}$$

as schematically shown in Fig. 5.10.
As apparent from Table 5.4, the denominator coefficients of all transfer functions depend on the controller parameters. That directly implies that in principle the pole locations can be explicitly set in certain limits by tuning the controller parameters.

Example: PT_1-type sensor with P-controller

The best accessible example to understand the influence of the closed-loop feedback structure on the sensor response time is the combination of a PT_1-type sensor with a P-controller. If one rearranges the respective transfer function from Table 5.4 one gets

$$G_{\mathrm{comp}}(s) = \frac{K^* k_{\mathrm{P}}}{1 + K^* k_{\mathrm{P}}} \cdot \frac{1}{1 + s\frac{T}{1+K^* k_{\mathrm{P}}}} = \frac{K_{\mathrm{comp}}}{1 + sT_{\mathrm{comp}}} \tag{5.32}$$

with

$$K_{\mathrm{comp}} = \frac{K^* k_{\mathrm{P}}}{1 + K^* k_{\mathrm{P}}}, \; T_{\mathrm{comp}} = \frac{T}{1 + K^* k_{\mathrm{P}}}. \tag{5.33}$$

From Eq. (5.32) one can observe that the closed-loop system is also a first-order system, however with a modified time constant T_{comp}. This time constant depends on the impressed time constant T of the open-loop sensor itself and inversely on the feedback factor. Thus, T is reduced to T_{comp} by the factor $(1/(K^* k_{\mathrm{P}}))$ and converges to zero for $K^* k_{\mathrm{P}} \to \infty$. The time-dependent sensor output signal for a unit step input of the open-loop PT_1-type sensor as well as the closed-loop sensor comprising a P-controller are shown in Fig. 5.11a for different feedback factors.
From Fig. 5.11a, one can observe the influence of different feedback factors on the sensor response time and on the static control deviation as discussed in Section 5.3.3.

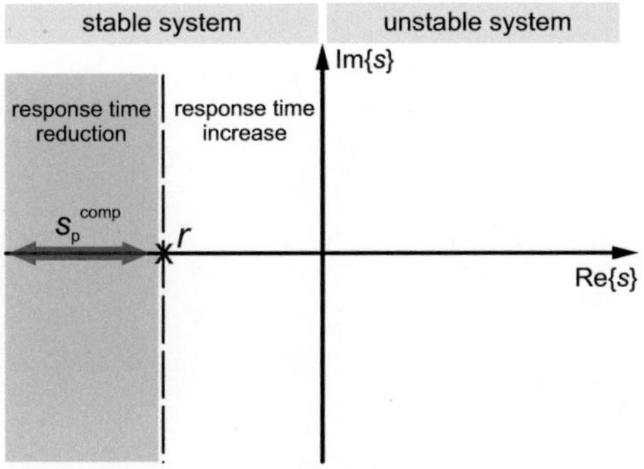

Figure 5.10: Representation of the complex GAUSS plane exhibiting the real part r of the dominant pole s_p^{defl} of the open-loop sensor and the LOH relative to this pole (gray region). The dominant pole s_p^{comp} of the closed-loop sensor has to be located in this half plane to enable a response time reduction according to Eq. (5.31).

It is noteworthy that even a unity feedback leads to a minimization of the closed-loop response time of 50 %, which can be seen in Fig. 5.11b. Here, the normalized closed-loop time constant is illustrated as a function of the feedback factor. Fig. 5.11c schematically illustrates the effect of the feedback structure on the pole location of the closed-loop sensor.

For the combination of more complex controller and sensor types, the transfer functions also become more complex. Hence, for denominator polynomials of increasing order (two or higher), the analytical determination of the poles becomes more demanding. At order two a case differentiation has to be made. For order three a general analytical solution is still possible but not trivial and may not lead to valuable practical conclusions. Therefore, to check the feasibility of a response time reduction and to verify the influence of the different controller types on the response time of the closed-loop system, we use a similar method as already used to proof system stability. This method verifies the requirements that have to be fulfilled, so that all poles of the closed-loop transfer function are placed in the LOH relatively to the dominant pole of the respective open-loop transfer function. This directly implies a response time reduction. The steps required to accomplish this are:

- the determination of the dominant poles of the open-loop transfer functions s_p^{defl} and its real parts $r = \Re\{s_p^{defl}\}$,
- the redevelopment of the denominator polynomials $D_{comp}(s)$ of the closed-loop transfer functions at point $s = r$ into the new polynomial $D_{comp}^r(s) = d_0^r + d_1^r(s-r) + \ldots + d_n^r(s-r)^n$ using the HORNER scheme,
- the evaluation of the coefficients $d_0^r \ \ldots \ d_n^r$ by application of the STODOLA condition and additional application of the ROUTH criterion (for $n > 2$) to

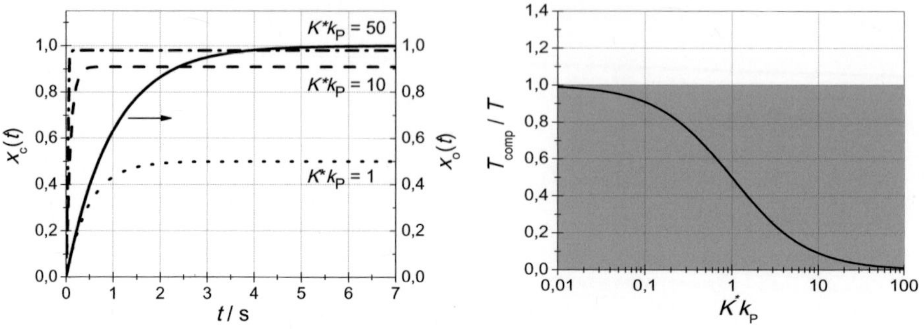

(a) (b)

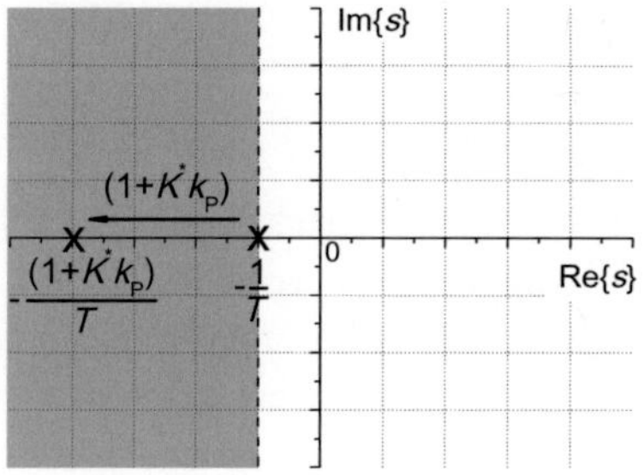

(c)

Figure 5.11: (a) Influence of the compensation method on the step response of an open-loop sensor (solid line) for $T = 1\,\mathrm{s}$ and a closed-loop sensor (interrupted lines) for different feedback factors, (b) normalized response time of a closed-loop sensor versus feedback factor; (c) schematic illustration of the pole-zero map indicating the pole shift due to the system's feedback structure for compensation. The gray region indicates where a response time reduction is achieved.

$D^{\mathrm{r}}_{\mathrm{comp}}$, to evaluate if all poles of the closed-loop transfer functions are placed left of $s = r$ (cp. gray region in Figs. 5.10 and 5.11c),

- the extraction of the controller parameters required to meet these conditions.

A detailed example of this procedure can be found in Appendix B.3. The dominant time constants of the open-loop sensor as well as the parameter requirements for a response time reduction for all sensor controller combinations are proposed in Table 5.10. From this table it can be seen that the application of the compensation method can lead to a decisive response time reduction. For a PT_1-type sensor, the requirements are clear and simple. For higher order sensor types, the requirements become more complex. However, in general, it can be observed that always only one controller parameter depends on the parameters of the open-loop sensor or has to be simply positive. The remaining controller parameters depend on the respective

other controller parameters as well and cannot be chosen individually. This can be exemplarily seen for the closed-loop time constant of a PT_1-type sensor in combination with a PI-controller in Fig. 5.12.
Here, the requirements from Table 5.10 are illustrated. It can be observed that a response time reduction is only possible if the feedback factor amounts to $K^* k_P > 1$ and if the integral time T_I is smaller than the time constant T of the open-loop sensor. The smaller the integral time is, the better the achieved response time reduction is. However, for every $K^* k_P$, a T_I exists where the response time becomes minimal. These minimum points are indicated by the locus (small-dotted line). A further minimization of T_I will only lead to an under-damped oscillatory system but not to a further response time reduction.

For the PT_2-type and the PT_2-D-type sensor two different cases according to the pole location were considered: (1) $\varphi = T_2/T_1 \leq 2$ (real double pole or a complex conjugate pole pair), (2) $\varphi = T_2/T_1 > 2$ (two real poles). Note that the combination of a PT_2-type sensor, which is critically-damped ($\varphi = 2$) or under-damped ($\varphi < 2$), with a P- or PI-controller will never lead to a response time reduction. Only the response time of over-damped systems ($\varphi > 2$) can be reduced using these controller types. However, the application of a PID-controller overcomes this limitation. Due to the dependence of all the specific requirements of all the sensor controller combinations on many parameters, a graphical illustration is not possible. Therefore, the controller parameters enabling a response time reduction have to be calculated for a specific problem. Subsequently, it has to be checked that the obtained parameter range is a subset of the parameter range for a stable minimum-phase system.

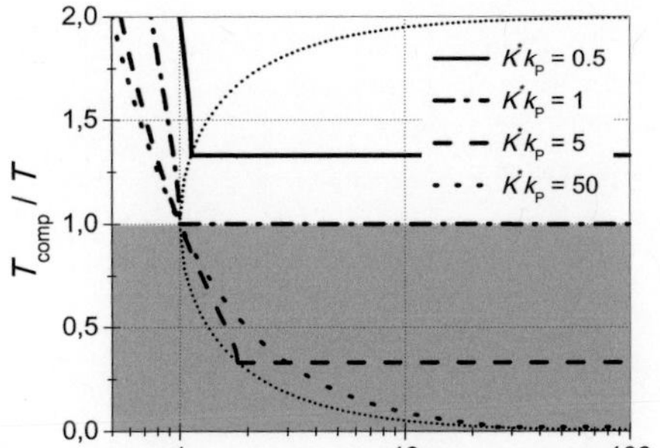

Figure 5.12: Normalized response time of the closed-loop PT_1-type sensor using a PI-controller versus integral time. The gray region indicates where a response time reduction is achieved.

Table 5.10: Dominant time constants $|1/r|$ of the three different sensor types as well as the requirements (parameter range) for a response time reduction of the closed-loop sensor for all considered sensor-controller combinations.

Open-loop sensor		Closed-loop sensor		
Sensor type	Dominant time constant $\|\mathbf{1/r}\|$	P	PI	PID
I: PT_1	T	$k_P > 0$	$k_P > \frac{1}{K^*}$ $k_I > \frac{k_P}{T}$	$k_P > \frac{1}{K^*} + \frac{2k_D}{T}$ $k_I > \frac{k_P}{T} - \frac{k_D}{T^2}$ [a] $k_D > -\frac{T}{K^*}$ [b]
II: PT_2				
(1) $\varphi \leq 2$:	$\frac{2T_1}{\varphi}$	–	–	$k_P > \frac{\varphi^2 - 4}{4K^*} + \frac{\varphi k_D}{T_1}$ $k_I > \frac{\varphi(4K^* k_P - \varphi^2 + 4)}{8K^* T_1} - \frac{\varphi^2 k_D}{4T_1^2}$ $k_D > \frac{\varphi T_1}{2K^*}$ $\frac{2K^*[k_I T_1^3 - k_D T_1(1 + K^* k_P) + \varphi K^* k_D^2]}{T_1(\varphi T_1 - 2K^* k_D)} > 0$ [c]
(2) $\varphi > 2$:	$\left\|\frac{2T_1}{-\varphi + \sqrt{\varphi^2 - 4}}\right\|$	$k_P > 0$	$k_P > \frac{\varphi\sqrt{\varphi^2 - 4} - \varphi^2 + 4}{2K^*}$ $k_I > \frac{k_P\left(\varphi - \sqrt{\varphi^2 - 4}\right)}{2T_1}$	$k_P > \frac{T_1(\varphi\sqrt{\varphi^2 - 4} - \varphi^2 + 4) + K^* k_D(2\varphi - 2\sqrt{\varphi^2 - 4})}{2K^* T_1}$ $k_I > \frac{k_P T_1(\varphi - \sqrt{\varphi^2 - 4}) + k_D(\varphi\sqrt{\varphi^2 - 4} - \varphi^2 + 2)}{2T_1^2}$

			$\varphi > \frac{3}{\sqrt{2}}$ $\frac{2K^* k_\mathrm{P}\sqrt{\varphi^2-4} - 2K^* k_\mathrm{I} T_1}{3\sqrt{\varphi^2-4}-\varphi} +$ $\frac{2\sqrt{\varphi^2-4}(\varphi^2-3) - 2\varphi^3 + 8\varphi}{3\sqrt{\varphi^2-4}-\varphi} > 0$ [c]	$k_\mathrm{D} > \frac{T_1(\varphi - 3\sqrt{\varphi^2-4})}{2K^*}$ $\frac{{K^*}^2 k_\mathrm{D}^2 (2\sqrt{\varphi^2-4} - 2\varphi)}{T_1(3T_1\sqrt{\varphi^2-4} - \varphi T_1 + 2K^* k_\mathrm{D})} +$ $\frac{K^* k_\mathrm{D}(4\varphi^2 - 4\varphi\sqrt{\varphi^2-4} - 14)T_1 - 2K^* k_\mathrm{I} T_1^3}{T_1(3T_1\sqrt{\varphi^2-4} - \varphi T_1 + 2K^* k_\mathrm{D})} +$ $\frac{T_1^2[\sqrt{\varphi^2-4}(2\varphi^2-6) + 8\varphi - 2\varphi^3]}{T_1(3T_1\sqrt{\varphi^2-4} - \varphi T_1 + 2K^* k_\mathrm{D})} +$ $\frac{2K^* k_\mathrm{P} T_1^2 \sqrt{\varphi^2-4} + 2K^* k_\mathrm{P} k_\mathrm{D} T_1}{T_1(3T_1\sqrt{\varphi^2-4} - \varphi T_1 + 2K^* k_\mathrm{D})} > 0$ [c]
III: PT_2-D				
(1) $\varphi \leq 2$:	$\frac{2T_1}{\varphi}$	N.A.	$k_\mathrm{P} > 0$ $k_\mathrm{I} > \frac{2\varphi K^* k_\mathrm{P} T_3 + T_1(\varphi^2-4)}{4K^* T_1 T_3}$	$k_\mathrm{P} > \frac{\varphi k_\mathrm{D}}{T_1}$ $k_\mathrm{I} > \frac{K^*(2\varphi k_\mathrm{P} T_1 - \varphi^2 k_\mathrm{D})T_3 + T_1^2(\varphi^2-4)}{4K^* T1^2 T_3}$ $k_\mathrm{D} > -\frac{T_1^2}{K^* T_3}$ [b]
(2) $\varphi > 2$:	$\left\lvert \frac{2T_1}{-\varphi + \sqrt{\varphi^2-4}} \right\rvert$	N.A.	$k_\mathrm{P} > -\frac{T_1\sqrt{\varphi^2-4}}{K^* T_3}$ [b] $k_\mathrm{I} > \frac{k_\mathrm{P}(\varphi - \sqrt{\varphi^2-4})}{2T_1}$	$k_\mathrm{P} > \frac{K^* k_\mathrm{D}(\varphi - \sqrt{\varphi^2-4})T_3 - T_1^2\sqrt{\varphi^2-4}}{K^* T_1 T_3}$ $k_\mathrm{I} > \frac{k_\mathrm{P}(\varphi - \sqrt{\varphi^2-4})T_1 + k_\mathrm{D}(\varphi\sqrt{\varphi^2-4} - \varphi^2 + 2)}{2T_1^2}$ $k_\mathrm{D} > -\frac{T_1^2}{K^* T_3}$ [b]

[a] This condition is only valid for positive k_I due to the stability requirement.
[b] This condition is automatically fulfilled due to the stability requirement.
[c] This condition results from the application of the ROUTH criterion.

b. Output signal characteristics

While the response time is determined by the poles of the transfer function, the zeros of the numerator influence the signal characteristics, i. e. overshoot or undershoot of the step response can occur. With the requirement of a minimum-phase system, we already excluded the possibility of an inverse sensor response, i. e. undershoot. However, the location of the zeros with respect to the pole location determines, whether the step response shows no maximum, one maximum or even several maxima before settling at the final steady state. For a proper sensor behavior, a step response which attains the final steady state as fast as possible, and hence shows no maximum, is convenient. This is only possible if the transfer function only contains real poles and zeros. The evaluation can be done best by rearranging the general transfer function from Eq. (5.12) in its pole-zero representation as

$$\begin{aligned} G_{\mathrm{comp}}(s) &= \frac{c_m\,(s-s_{01})\cdot(s-s_{02})\cdot\ldots\cdot(s-s_{0m})}{d_n\,(s-s_{\mathrm{p}1})\cdot(s-s_{\mathrm{p}2})\cdot\ldots\cdot(s-s_{\mathrm{p}n})} \\ &= \frac{c'_m\,(1+T_{\mathrm{N}1})\cdot(1+T_{\mathrm{N}2})\cdot\ldots\cdot(1+T_{\mathrm{N}m})}{d'_n\,(1+T_{\mathrm{D}1})\cdot(1+T_{\mathrm{D}2})\cdot\ldots\cdot(1+T_{\mathrm{D}m})}, \end{aligned} \tag{5.34}$$

with

$$s_{0\mu} = -\frac{1}{T_{\mathrm{N}\mu}},\quad s_{\mathrm{p}\nu} = -\frac{1}{T_{\mathrm{D}\nu}}. \tag{5.35}$$

and $\mu = 1, \ldots, m$, $\nu = 1, \ldots, n$.

For further considerations, we assume the time constants to be ordered by size as

$$\begin{aligned} T_{\mathrm{N}1} &\geq T_{\mathrm{N}2} \geq \ldots \geq T_{\mathrm{N}\mu}, \\ T_{\mathrm{D}1} &\geq T_{\mathrm{D}2} \geq \ldots \geq T_{\mathrm{D}\nu},\ . \end{aligned} \tag{5.36}$$

A step response without overshoot (no maximum) is possible, if all time constants are real, and every numerator time constant $T_{\mathrm{N}\mu}$ is smaller than the corresponding denominator time constant $T_{\mathrm{D}\nu}$.

Therefore, the controller parameters have to be chosen such that the above requirement is fulfilled. Consequently, a transfer function exhibiting only real poles and no zeros never shows an overshoot of its step response. Note that changing the numerator time constants also changes the denominator time constants, because they depend on each other via the controller parameters and cannot be changed individually.

5.4 Conclusions and summary

Based on a general sensor model, both the deflection and the compensation method have been described and compared in a general systematic fashion. A system-based approach was applied with regard to important figures of merit of sensor technology. In particular, the quasi-static sensor behavior, comprising systematic static variations of the sensitivity, long-term drift effects as well as the dynamic sensor behavior, i. e. response time and output signal characteristics, have been described. Three sensor types, i. e. PT_1-, PT_2-, and PT_2-D-type sensors, which cover most of the existing sensors, were analyzed concerning both measurement methods.

In conclusion, for a sensor operated with the deflection method the following applies:

- A static sensitivity change of the sensor due to an external interference quantity directly influences the sensor output signal.
- Quasi-static sensitivity changes, i. e. long-term drift effects, lead to a continuous infinite increase of the output signal, whereas an explicit assignment of the input signal to the output signal is not possible without additional stipulations.
- The response time is impressed by the sensor principle, the sensor material properties, and by the underlying physical phenomena, respectively. This directly reflects in the coefficients of the denominator polynomial of the respective transfer function.
- For the transfer functions studied, the output signal characteristics is in direct correlation with the response time of the sensor and, thus, just determined by the denominator coefficients.
- The system complexity is relatively low compared to sensors with force compensation.

For a sensor operated with the compensation method, the following conclusions can be drawn:

- Such sensors constitute a feedback system as known from feedback loops in automatic control.
- Different from feedback loops in automatic control, the measuring signal is the compensation quantity, i. e. the output signal of a controller/actuator unit.
- Like for every feedback loop system, system stability plays a decisive role and has to be assured for all sensor operation conditions.

- For the studied closed-loop sensor transfer functions, stability and additional minimum-phase behavior are assured for the set of positive controller parameters.
- Static sensitivity changes of the core sensor element due to an external interference quantity are suppressed by the static gain of the feedback elements. Thus, the influence on the output signal is only marginal and for some sensor controller combinations even zero.
- Sensors exhibiting long-term drift effects can be stabilized, so that a constant output signal is achieved at the steady state. This is even possible with a simple unity feedback. P- and PI-controllers have been found to be most suitable for drift suppression, whereas PID-controllers are less adequate.
- The poles of the closed-loop sensor transfer function can be tuned by the controller parameters. In principle, the response time can be decisively reduced compared to an open-loop sensor. The requirements for a response time reduction have been calculated for all sensor-controller combinations. However, for a PT_2-type sensor with a real double pole or a complex conjugate pole pair, a response time reduction is not possible with a P- or PI-controller.
- The output signal characteristics depends on the location of the zeros with respect to the location of the poles. However, both zeros and poles are a function of the controller parameters and cannot be set separately.
- To determine the optimized controller parameters, a detailed model of the core sensor element and all possible peripheral devices must exist, which makes the sensor design more demanding.
- Compared to the deflection method, the application of the compensation method results in a higher system complexity due to the necessity of additional devices like controller and actuator.

6 Closed-loop hydrogel-based sensor [1]

In this chapter, the compensation method, particularly applied to hydrogel-based pH sensors, is described. After the introduction of the general structure and principle of a pressure-compensated hydrogel sensor, the design, materials, and fabrication techniques of the sensor element, containing the hydrogel sensing layer, are discussed. Subsequently, the integration of the sensor element into the closed-loop system structure as well as the set-up, characterization and modeling of the individual system components is described. In particular, a macroscopic rheological gray-box approach is used to describe the hydrogel behavior. The obtained model structure, together with experimentally derived models for the remaining components, is implemented in MATLAB/Simulink and a simulation of the closed-loop sensor is performed to obtain a controller parameter range where the system is stable and shows the desired system behavior. As a result of the simulation, selected controller parameters are applied to the closed-loop sensor and the measurement results are presented and discussed.

6.1 Sensor concept

6.1.1 General sensor principle and sensor structure

According to the general structure of a compensation circuit, as shown in Fig. 5.6, the closed-loop hydrogel sensor with pressure compensation is basically built up on four main components: (i) the sensor element containing the sensitive hydrogel layer, (ii) a measuring unit, (iii) a controller, and (iv) an actuator for supplying the compensation pressure (Fig. 6.1a). What is called here the sensor element is a device part that encloses the stimuli-sensitive hydrogel layer in a pre-defined constant volume, i. e. the hydrogel is mechanically constrained from all sides. However, one side of this enclosure must allow supply of the analyte solution to be measured but has to be mechanically rigid as well as impermeable for the hydrogel even under high pressure. Furthermore, one side has to be able to become displaced to transfer the compensation pressure supplied by the actuator to the hydrogel in order to maintain a constant hydrogel volume. The hydrogel's attempt to change its volume – when the analyte concentration is changed – is continuously detected by a measuring unit. The resulting output signal of the measuring unit is fed to a controller. The controller

[1]Parts of this chapter are based on a publication by the author in "*IEEE Sensors Journal*" [195].

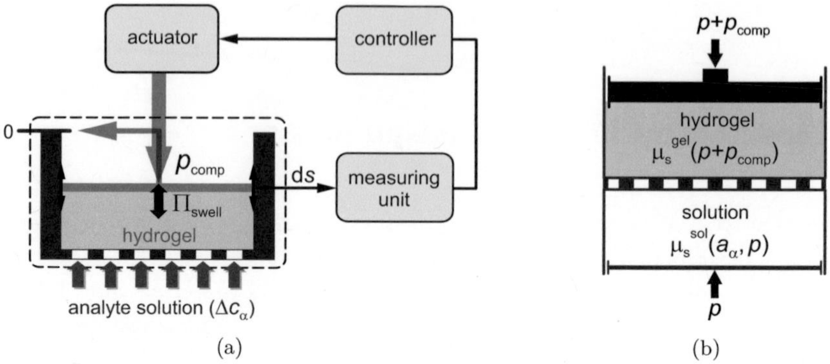

Figure 6.1: (a) General set-up and its components of a closed-loop hydrogel-based sensor with pressure compensation, (b) simplified model of the sensor element with applied compensation pressure comprising a hydrogel and a solution compartment separated by a membrane which is permeable for the solution.

provides a defined signal to the actuator, according to the respective control law, which again delivers the compensation pressure p_{comp} in such a way that p_{swell} is fully compensated and the null-indicator remains in quasi-null-position.

In contrast to the description of hydrogel swelling in Section 2.1, where isobar conditions were presumed, the hydrogel inside the closed-loop sensor undergoes quasi-isochoric state changes. At the steady state, a constant hydrogel volume is enforced by application of p_{comp} which exceeds the surrounding pressure p (e. g. atmospheric pressure) (Fig. 6.1b). The resulting equilibrium condition

$$\mu_{\text{s}}^{\text{sol}}(a_\alpha, p) = \mu_{\text{s}}^{\text{gel}}(p + p_{\text{comp}}) \tag{6.1}$$

holds under the assumption that only the pure solvent exists in the gel phase and the solution phase contains the analyte of activity a_α. Here, $\mu_{\text{s}}^{\text{sol}}$, $\mu_{\text{s}}^{\text{gel}}$ are the chemical potential of the solvent in the solution and in the gel, respectively. The left hand and the right hand term in Eq. (6.1) can be rewritten as

$$\mu_{\text{s}}^{\text{sol}}(a_\alpha, p) = \mu_{\text{s}}^{\text{gel}}(p) + RT \cdot \ln a_\alpha \tag{6.2}$$

and

$$\mu_{\text{s}}^{\text{gel}}(p + p_{\text{comp}}) = \mu_{\text{s}}^{\text{gel}}(p) + \int_p^{p+p_{\text{comp}}} V_{\text{s,m}} \text{d}p, {}^2 \tag{6.3}$$

respectively. Application of Eqs. (6.2) and (6.3) to Eq. (6.1) and transposition to p_{comp} yields

[2] $V_{\text{s,m}}$ is assumed to be constant in the considered pressure range.

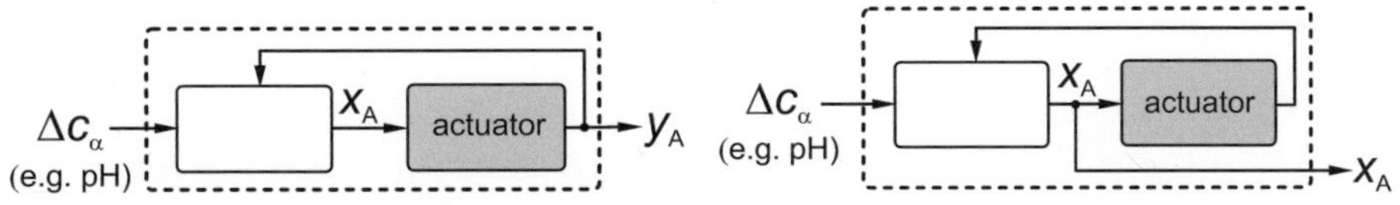

Figure 6.2: Possible sensor structure of a pressure-compensated chemical sensor: (a) The output signal of the actuator y_A is directly measured and a transfer characteristic $y_A = f(\Delta c_\alpha)$ is obtained. (b) The input signal of the actuator x_A is directly measured and a transfer characteristic $x_A = f(\Delta c_\alpha)$ is obtained.

$$p_{\text{comp}} = \frac{RT}{V_{\text{s,m}}} \ln a_\alpha = \frac{RT}{V_{\text{s,m}}} \ln (\gamma \cdot c_\alpha) = -\Pi_{\text{swell}}. \tag{6.4}$$

Thus, in the steady state, the applied compensation pressure equals the osmotic swelling pressure and, thus, p_{comp} counteracts Π_{swell} (negative sign). Hence, the compensation pressure is a direct measure of the analyte concentration to be measured.

If the output signal y_A of the actuator (Fig. 6.2a) is not directly accessible or if a measurement of this quantity is too complicated, especially with regard to a miniaturized actuator, the input signal x_A of the actuator can alternatively be measured and evaluated (Fig. 6.2b) leading to a different transfer function.

6.1.2 Theoretical implications for hydrogel sensors in closed-loop configuration

As described in Section 2.1.2, the expansion of the polymer network in the measuring solution is the main factor influencing the time constant of an open-loop sensor due to cooperative diffusion. Furthermore, the visco-hyperelastic material properties lead to a long-term drift of the time-dependent sensor output signal arising from a slow linear increase in hydrogel volume for long measuring times. Thus, ideally, a quasi-constant hydrogel volume achieved by a closed-loop feedback would imply:

- The cooperative movement of the polymer chains in the measuring solution can be avoided leading to the suppression of the cooperative diffusion process.
- The time constant due to cooperative diffusion can be decreased and the sensor response time should be only influenced by the dominant time constant of the feedback path.
- Linear long-term drift and relaxation effects can be suppressed.
- Hysteretic sensor behavior resulting from plastic deformation of the gel network under large extension can be decreased.

6.1.3 Design specifications

From the above considerations concerning the basic structure and the operational principle of the hydrogel sensor with pressure compensation, specifications for the sensor design are derived in Table 6.1. Here, the targeted properties are essential for a proper sensor operation whereas the other design goals will support the application in different technical environments.

Table 6.1: Design specifications for the closed-loop hydrogel-based sensor with pressure compensation.

Sensor component	Targeted properties	Other design goals
Sensor element	• Mechanically rigid • Permeable for the measuring solution • Impermeable for the hydrogel network	• Potential for miniaturization • Fabrication by micromachining • Applicability to different actuator principles
Measuring unit	• High sensitivity • Small response time	• Spatial separation of the electronic components from the measuring solution and the hydrogel
Controller		• Linear standard controller
Actuator	• Precise pressure supply in a wide pressure range • Small response time	

6.2 Construction of the sensor element

6.2.1 Structural set-up

According to the design specifications in Table 6.1, the core part of the sensor is fabricated by conventional micromachining techniques to enable scalability for specific applications as well as further miniaturization. Therefore, due to its excellent properties regarding mechanical stability and the possibility of surface modification as well as the scalability of the size of volumetric and surface features by standard microsystem fabrication techniques, a silicon-based sensor platform is proposed here (Fig. 6.3).

The rigid silicon backplate is the self-supporting basis of the device and contains a square cavity in its center. This cavity provides a defined constant volume and contains the sensitive hydrogel layer. The bottom of the cavity is connected to the backside of the backplate by cylindrical microholes allowing supply of the measuring solution from the backside of the plate via a 3×3 microhole array (Fig. 6.4). The diameter and the pitch of the microholes are chosen to enable direct influence of the measuring solution on the hydrogel layer, i. e.

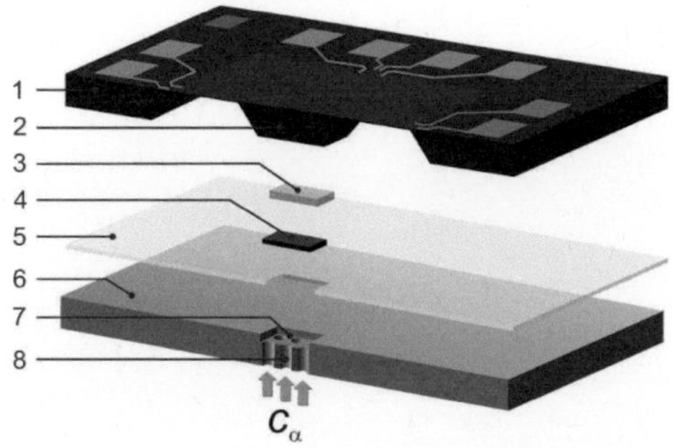

Figure 6.3: Sensor element in a half-sectional exploded view with its individual device components; 1 silicon micromachined piezoresistive pressure sensor chip with 2 center boss structure, 3 pH-sensitive hydrogel layer, 4 nanoporous membrane, 5 intermediate bonding layer, 6 rigid silicon backplate, 7 square cavity, 8 cylindrical fluidic microholes.

- preferably high porosity of the microhole array,
- mechanical stability and rigidity,
- reliable processing of the microhole arrangement by anisotropic etching.

The porosity of the cavity, i. e. the ratio of the open area to the total area of the cavity, is 38 % (Table 6.2). This is comparable to the design by ORTHNER *et al.* where holes with a diameter in the range between 100 µm and 175 µm with a pitch of 200 µm have been used, resulting in a porosity between 28 % and 41 % [12].

A permeable membrane which is located between the hydrogel layer and the microholes in a sandwich-like structure prevents evasion of the hydrogel network into the microholes during the swelling process (Fig. 6.4a). This is especially important to assure a constant volume of the hydrogel network and to attain a high swelling pressure which would otherwise be reduced by hydrogel expansion into the open microholes. However, the solution transport from the sensor backside to the hydrogel must not be hindered by the membrane. From the top, the cavity is capped by a square center boss which is a monolithic part of a micromachined silicon pressure sensor chip. The pressure sensor chip is bonded to a backplate by a soft-bonding technique using an intermediate bonding layer. To assure reliable closure of the cavity volume, the cavity dimensions have to be smaller than the bottom face of the boss structure, such that a sufficient overlap is achieved (see Fig. 6.4).

In this sensor design, the pressure sensor chip combines two functionalities:

- The boss structure and the thin flexure plate act as flexible actuator components which directly act on the hydrogel structure in order to maintain a constant volume.

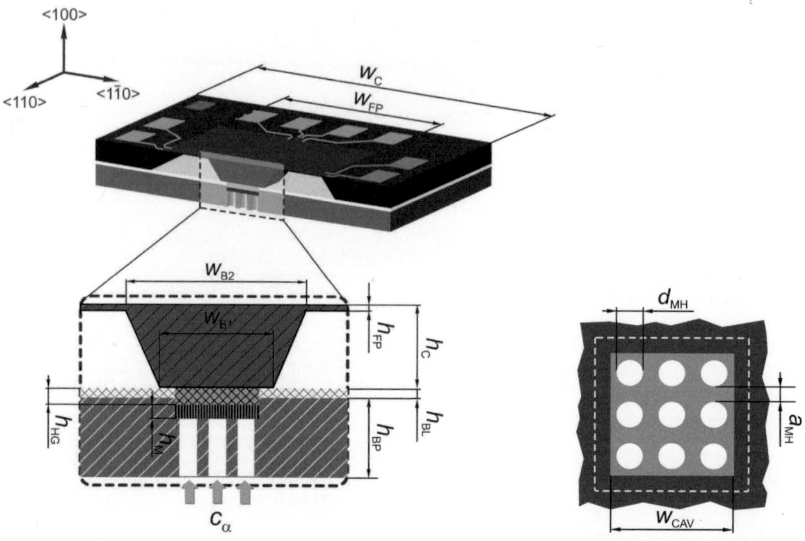

(a) (b)

Figure 6.4: Sensor element of the closed-loop hydrogel sensor: (a) assembled device structure and detailed cross-sectional view of its center part, (b) top view of the cavity for enclosure of the hydrogel layer, the light gray interrupted line indicates the outer dimensions of the bottom face of the center boss structure; the geometric dimensions indicated here are quantitatively summarized in Table 6.2. h height, w width, d diameter, a distance. Indices: C sensor chip, B center boss, FP flexure plate, BL bonding layer, M porous membrane, HG hydrogel, BP backplate, MH microhole, CAV cavity.

- The pressure sensor chip is part of the measuring unit and works as pressure transducer detecting the hydrogel's attempt to change its volume via bending of the flexure plate.

In this way, parts of the actuator and the measuring unit are monolithically integrated in the same structure which enables compact packaging. Furthermore, a strict spatial separation between the (sometimes aggressive) analyte solution at the backside and the electronic components at the frontside of the pressure sensor chip is achieved. The fluid components only interact with the backplate and the bottom face of the boss structure via the hydrated hydrogel, whereas the top of the pressure sensor chip, containing sensitive electronic components, is completely protected from it. This enables an excellent feature regarding the long-term stability of the sensor element. Table 6.2 lists the geometric parameters of the sensor element used for experimental studies.

Table 6.2: Geometric parameters of the sensor element used for experimental studies (cp. Figs. 6.3 and 6.4).

Component	Parameter	Symbol	Value
1,2 Pressure sensor chip	Chip dimension	w_C	6500 µm
with boss structure	Chip thickness	h_C	400 µm
	Dimension of flexure plate	w_{FP}	3500 µm
	Thickness of flexure plate	h_{FP}	20 µm
	Dimension of boss (bottom face)	w_{B1}	1000 µm
	Dimension of boss (top face)	w_{B2}	1500 µm
3 Hydrogel layer	Hydrogel thickness	h_{HG}	50 µm
4 Nanoporous membrane	Membrane thickness	h_M	60 µm
5 Intermediate bonding layer	Layer thickness	h_{BL}	50 µm
6 Backplate	Height of backplate	h_{BP}	400 µm
7 Square cavity	Cavity dimension	w_{CAV}	700 µm
8 Fluidic microholes	Hole diameter	d_{MC}	150 µm
	Hole distance	a_{MC}	100 µm

6.2.2 Fabrication of the sensor element components

a. Piezoresistive pressure sensor chip with center boss structure

The piezoresistive pressure sensor chips are an in-house fabrication by the Physikalisch-Technische Bundesanstalt (PTB, Braunschweig, Germany) (Fig. 6.5). The thinning of the flexible part between the thick outer rim and the center boss was done using anisotropic wet etching of a (100)-oriented silicon wafer. The stiffened center part is fabricated by utilization of an etching mask and has the shape of a truncated pyramid with a characteristic angle of the side walls of 54.7 °. This fabrication technique is simple and cost-efficient. However, the top face of the boss structure, which is in contact with the flexure plate, has a larger area than the bottom face ($w_{B2} > w_{B1}$) which is the effective area being in contact with the hydrogel (cp. Table 6.2 and Figs. 6.5a,c). Thus, the stiffening of the flexure plate is more pronounced compared to a boss structure with ideally vertical side walls, leading to a reduced sensitivity. Four piezoresistors, directed along the <110>-direction near the surface of the flexure plate, are located at the middle of each outer edge of the flexure plate, respectively (cp. red dashed lines in Fig. 6.5b). Thus, in contrast to most conventional pressure transducers, exhibiting one WHEATSTONE bridge circuit, four WHEATSTONE bridge circuits, containing four piezoresistors each, are featured by this transducer design. Ideally, this leads to a four times higher transducer sensitivity, if all terminals are

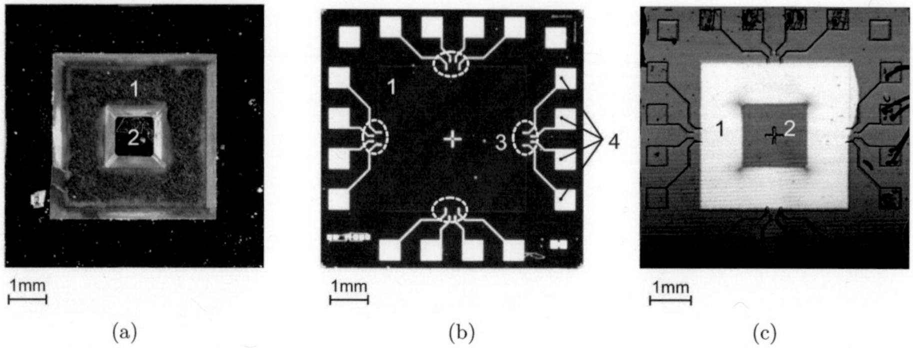

Figure 6.5: Micromachined silicon piezoresistive pressure transducer with center boss structure: (a) view from the transducer backside into the etched cavity, (b) top view onto the chip surface containing the electrical paths, piezoresistors, and metallic bond pads; the four WHEATSTONE bridge circuits containing four piezoresistors each are indicated by the dashed lines, (c) acoustic microscope image of the sensor surface; 1 silicon flexure plate, 2 center boss structure, 3 four piezoresistors in WHEATSTONE bridge configuration, 4 metallic bond pads.

used and the respective output voltages are summed up. The resistance of the piezoresistors was determined to be $(1.55 \pm 0.01)\,\mathrm{k\Omega}$ and a supply voltage of 5 V was used.

b. Sensor backplate with microhole array

To allow the investigation of the microhole array parameters on the sensor performance, a sensor backplate layout containing chips with different microhole diameters and different distance between microholes was used. For simplicity reasons, the entire area of the sensor backplate has been perforated by microholes even if only the center part (3×3 hole array, Fig. 6.4) was used for coming into contact with the hydrogel element.

The main technological steps of the fabrication process (performed at IHM, TU Dresden) can be outlined as (Fig. 6.6):

1. Spin-coating of a thick-film photoresist (AZ 4562, 9 µm) on both sides of a silicon wafer, UV photo-patterning of the wafer top side (mercury vapor lamp, $\lambda = 365\,\mathrm{nm} \ldots 436\,\mathrm{nm}$, $t_{\mathrm{exp}} = 16\,\mathrm{s}$) by contact lithography using a foil mask to establish a structured etch stop layer.

2. Deep reactive ion etching (DRIE) by application of the Bosch-process [196] to accomplish an array of open cylindrical microholes normal to the wafer surface (passivation step by polymer coating: 85 sccm C_4F_8, 21 mTorr, $P_{\mathrm{ICP}} = 600\,\mathrm{W}$,

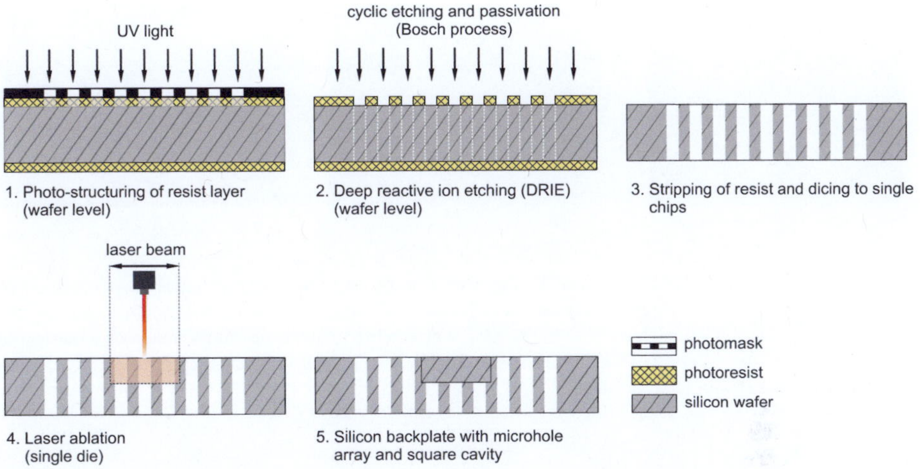

Figure 6.6: Fabrication of the rigid backplate with microhole array and square cavity (not drawn to scale).

8 s; silicon etching: 130 sccm SF_6, 40 mTorr, $P_{ICP} = 600$ W, $P_{HF-Elektrode} = 12$ W, 12 s; total number of cycles ca. 270, process time 90 min) (Figs. 6.7b,c).

3. Stripping of resist layer (subsequent washing in aceton, isopropyl alcohol and deionized water). Dicing of the wafer into single chips of (6.5×6.5) mm^2 by wafer sawing (Fig. 6.7a).

4. Processing of the square cavity by laser ablation (Ytterbium-fibre laser, $\lambda = 1.06$ µm, $P_{max} = 20$ W, microCUT, 3D-Micromac). The backplate is placed on a x-y-table and moved in a two-dimensional pattern below the laser spot such that an area of (700×700) μm^2 is exposed with a positioning and repeat accuracy of 10 µm, respectively. Since laser ablation is a maskless thinning technique, it has been favored to realize adjustable cavities.

c. pH-sensitive hydrogel material

A cross-linked polymer blend of PVA and PAA was used as pH-sensitive material. This material is one of the best-studied polymer systems due to its promising applications in biomedical systems [197, 198]. PVA/PAA hydrogels (Fig. 6.8) exhibit a mechanically stable structure and facilitate good processability as well as relatively simple integration into the microsensor. PVA provides a mechanically stable matrix with good long-term temperature and pH stability. PAA is anionic containing carboxyl groups and, therefore, constitutes the pH-sensitive component that accounts for the

(a) (b) (c)

Figure 6.7: Fabrication of the silicon backplate: (a) detail of an optical microscope imgage of a 4" wafer containing the processed $(6.5 \times 6.5)\,\mathrm{mm}^2$ Si backplates with microholes of different diameter and different inter-channel distance after the sawing process, (b) SEM image of a backplate containing the array of microholes, (c) cross-sectional view along a line of perpendicular microholes, showing the open channel structure.

polyelectrolytic character of the gel. Appendix A.3 summarizes both the properties and the synthesis route of PVA/PAA hydrogels.

To form PVA/PAA hydrogel layers, the polymer solution, exhibiting a honey-like viscosity, was spin-coated onto a polished 2" silicon wafer. The spin-coating parameters, to achieve a defined layer thickness, differ with respect to the water content of the polymer solution. The water content and, thus, the viscosity change due to evaporation. The un-cross-linked polymer film was dried for about 48 hours at room temperature. The resulting dry layer was carefully peeled off, placed on a PTFE disc and, subsequently, isothermally annealed in an oven at 130 °C for 20 minutes to induce physical cross-linking of both base polymers resulting in a swellable but insoluble network structure. The placement of the layer on the PTFE carrier during the annealing step avoids strong surface binding and, hence, rupture of the film when it is removed. The cross-linking regime is optimized regarding sensor-actuator functionality and mechanical stability [199]. As a result, transparent films with high mechanical stability and a thickness of about 50 µm have been obtained. These films were manually cut into $(700{\times}700)\,\mu\mathrm{m}^2$ pieces using a scalpel.

(a) (b)

Figure 6.8: Chemical structure of the hydrogel components: (a) poly(vinyl alcohol) (PVA), (b) poly(acrylic acid) (PAA).

d. Nanoporous membrane

The nanoporous membrane acts as a barrier layer between the hydrogel matrix and the microholes. This kind of membranes are often used in applications like filtration, separation as well as sorting, sensing, and release of biological molecules because the membrane properties can be tailored for specific applications [200]. According to the targeted sensor operation, the membrane material should feature

- a small pore size in the sub-µm range with uniform pore structure to form a hierarchical structure in conjunction with the microholes inside the backplate and to avoid hydrogel swelling into the microholes,
- high porosity,
- narrow pore size distribution,
- hydrophilicity to assure an unhindered transport of the analyte solution to the hydrogel,
- mechanical strength to withstand high normal forces,
- chemical and thermal stability,
- good processability.

Table 6.3 lists typical types of nanoporous membranes and selected properties. Inorganic membranes are most suited to fulfill the above mentioned requirements, especially because of their good mechanical and chemical stability as well as their high pore density. In particular, anodized aluminium oxide (AAO), fabricated by an electrochemical etching process, shows excellent hydrophilicity as well as an ordered

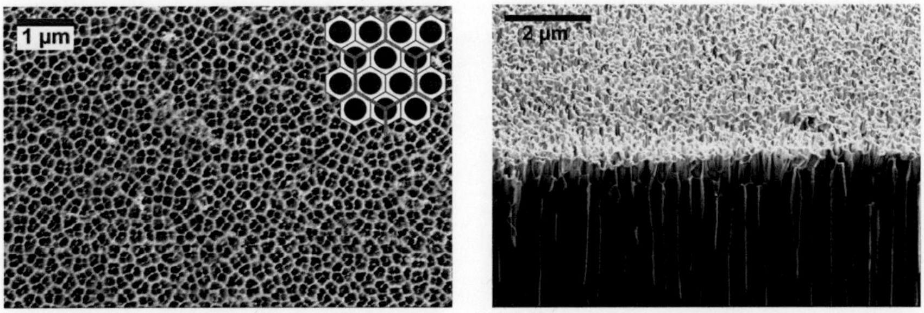

(a) (b)

Figure 6.9: SEM image of an Anapore™ nanoporous membrane with a pore diameter of about 200 nm: (a) top view on the membrane surface and its ideal hexagonal lattice structure when fabricated by anodization, (b) cross-sectional view of a cleaved sample showing the straight-ordered pore structure.

Table 6.3: Selected types of nanoporous membranes and their properties according to [200].

Type	Stability Mechanical/ chemical	Material	Fabrication method	Pore properties Size/distribution/ density/morphology
Inorganic	+/++	Anodic Al_2O_3, SiO_2	Anodization process	> 10 nm/narrow/ 10^{10} cm^{-2}/ordered, straight
		Ceramics (Al_2O_3, SiO_2, ZrO_2, TiO_2)	Powder sintering	1...50 nm/wide/ –/tortuous
			Sol gel	> 2 nm/narrow/ 10^{10} cm^{-2}/tortuous
			Micromachining	> 10 nm/narrow/ 10^{9} cm^{-2}/ordered
Organic	o/–o	Polymers (PC, PE)	Ion-track etching	> 20 nm/narrow/ 10^{8} cm^{-2}/ordered
		Polymers (PC, PE, PET, PS)	Phase separation	> 100 nm/medium/ high/tortuous
			Lithography	–
Organic/ inorganic compounds		Taylorable		Taylorable

PC polycarbonate, PE polyether, PET polyethylene terephthalate, PS polystyrene;
(++) very good , (+) good, (o) fair, (–o) poor-fair

structure of cylindrical pores, ideally arranged in a hexagonal lattice structure and aligned normal to the membrane surface [201]. To take advantage of these properties, Anopore™ inorganic membranes (Whatman, Springfield Mill, UK) with an average pore diameter of 200 nm have been used for this work. Their surface and pore

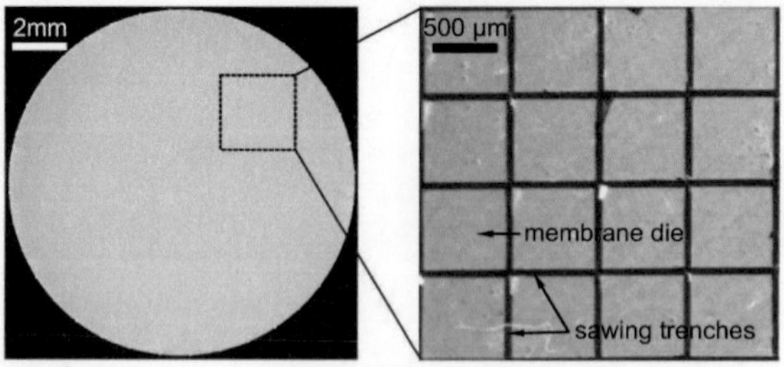

Figure 6.10: Optical microscope image of a plane 13 mm aluminum nanoporous membrane disc (Anapore) and the square membrane pieces after dicing. The square area indicates the proportions of the diced area with respect to the size of the entire membrane disc.

morphology is shown in Fig. 6.9. It is supplied in circular sheets with a diameter of 13 mm and a thickness of 60 µm and exhibits a porosity of 25 % to 50 %. It can be reliably exposed to temperatures of up to 400 °C and shows a burst pressure between 445 kPa and 755 kPa. AAO processing is compatible with microfabrication and allows precise engineering of the nanoscale morphology. However, the brittle character of this material complicates the fabrication of membranes with arbitrary shape and size and requires more sophisticated cutting techniques. The mounting of the circular membrane to dicing tape and subsequent cutting by a dicing saw to (700×700) µm^2 pieces has been found to work excellent with good reproducibility and exact match of the desired dimensions (Fig. 6.10).

e. Intermediate bonding layer

For sensor assembly, the rigid silicon backplate and the piezoresistive pressure transducer have to be bonded together with high reliability. In principle, conventional wafer bonding techniques like anodic bonding, direct wafer bonding (silicon fusion bonding), and intermediate layer bonding (eutectic bonding, glass-frit bonding) necessitate temperatures in the range from 300 °C up to 1200 °C to establish reliable bonding [202, 203]. In case of silicon fusion bonding, the two silicon surfaces have to be extremely clean and smooth what is demanding with respect to material surface preparation and necessitates a cleanroom environment.
Since the organic PVA/PAA material is used as sensing material, high temperatures during the bonding process would cause additionally cross-linking which would directly alter the sensor properties. Therefore, bonding should be performed at temperatures as close as possible to room temperature. Using poly(dimethylsiloxane) (PDMS) as intermediate layer material for intermediate layer bonding, bonding could be performed at room temperature. PDMS is well-known as standard material for polymer-based microfluidic applications due to its outstanding properties which can be advantageously used for the assembly process. For our experiments we used Sylgard 184 (Dow Corning, Midland, USA) which is a two-component system comprising

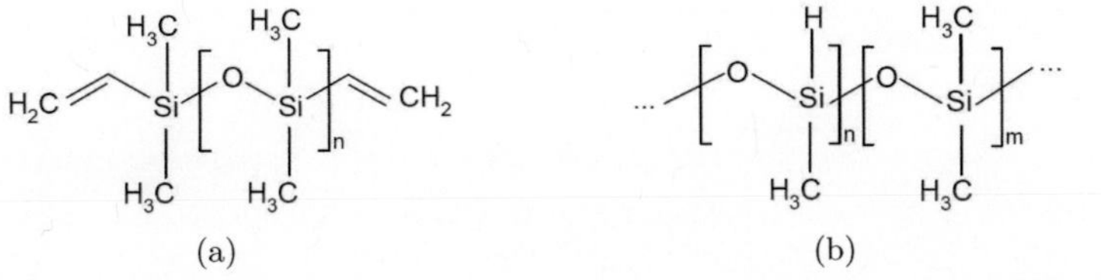

Figure 6.11: Basic components of Sylgard 184 elastomer (Dow Corning): (a) dimethylvinyl-terminated siloxane (n≈ 60), (b) dimethyl methylhydrogen siloxane (n≈ 10). During the cross-linking reaction, the hydrogen atoms of component (b) react with the vinyl groups of component (a).

dimethylvinyl-terminated siloxane as base material and dimethyl methylhydrogen siloxane as crosslinker (Fig. 6.11). Additionally, the cross-linking solution contains a platinum catalyst and a catalyst inhibitor [204]. The cross-linking reaction can even take place at room temperature but elevated temperatures (up to 150 °C) accelerate the cross-linking reaction and yield cross-linking times of only a few minutes. Advantageously, the mechanical properties can be tailored by the mixing ratio of both components and the cross-linking temperature. The un-cross-linked solution is clear with a honey-like viscosity and is thus well-suited for spin-coating of layers with defined thickness. Cross-linked PDMS shows a highly hydrophobic surface due to its small surface energy, good temperature stability up to 200 °C, and very good mechanical properties like high elasticity, very low stress relaxation and creep, and high compressibility [205, 206].

The utilization of PDMS as bonding layer is possible by a plasma bonding process and is primarily used to bond it to glass, silicon or other PDMS surfaces [207, 208]. The surface modification of the involved materials by oxygen plasma creates reactive hydroxyl groups on each surface and enables permanent reliable bonding by means of a condensation reaction when the materials are brought into contact (Fig. 6.12). The bonding process can be carried out at room temperature and does not require specifically smooth or clean surfaces of the materials involved. The thickness of the bonding layer can be precisely adjusted, e. g. by spin coating or dip coating.

To fabricate the PDMS intermediate layer, the base component and the curing agent have been mixed using a weight ratio of 7.5:1. The solution was constantly stirred to enable a homogeneous distribution of the curing agent. Subsequently, the solution was degassed under vacuum for 20 minutes in order to remove air bubbles and then spin-coated onto a polished 2" silicon support wafer. The coated wafer was annealed for 10 min at 95 °C on a hot plate to enable a cross-linked, mechanically stable layer.

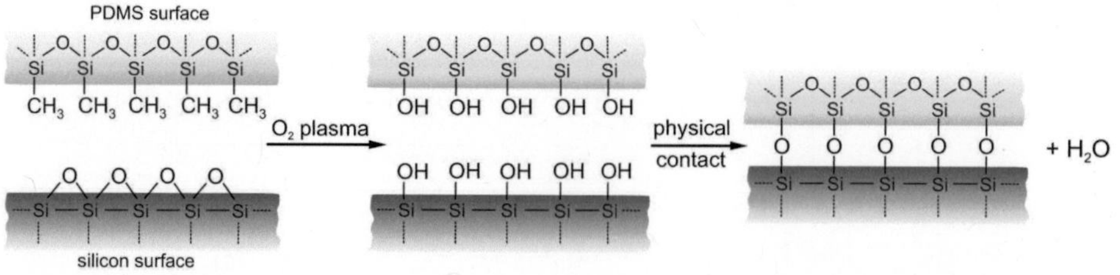

Figure 6.12: Generation of reactive hydroxyl groups on both silicon and PDMS surfaces by oxygen plasma treatment and covalent bonding by means of a condensation reaction occuring when both surfaces are brought into physical contact.

6.2.3 Assembly and packaging

Fig. 6.13 shows the assembly process of the individual sensor components by the following steps:

1. Surface modification of the PDMS layer and the top side of the backplate by oxygen plasma (400 W, 20 Pa, 70 s).

2. Physical contact of both surfaces under load ($p_{\text{contact}} = 49\,\text{kPa}$) for several hours to enable covalent bonding. The surfaces should be brought into contact within 1 min because the density of hydroxyl surface groups is decreasing with time and the surface becomes hydrophobic due to diffusion of un-cross-linked oligomers to the surface.

3. Support wafer detachment by means of a slim blade for initial separation of the material interfaces and subsequent gentle lift-off of the support wafer. Cut-out of a square area of (700×700) µm^2 – matching the cavity dimensions – in the center of the PDMS layer using a scalpel.

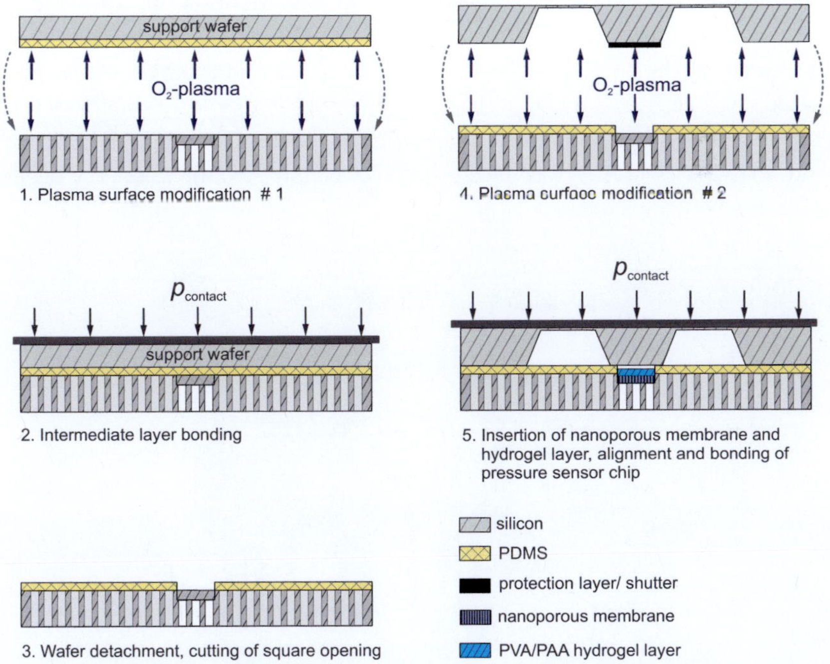

Figure 6.13: Assembly of the sensor element.

4. Surface modification of the PDMS surface and the backside of the pressure sensor chip by oxygen plasma (same parameters as in step #1). During this step, the bottom face of the boss structure has to be covered to prevent surface modification by oxygen plasma. Plasma protection can be established by simply covering the area with a thin ink layer (e. g. from a permanent marker) or by using a shutter which is placed on top of the boss structure.

5. Insertion of the nanoporous membrane die and the hydrogel layer, respectively, into the cavity using vacuum tweezers (assembly of the hydrogel after plasma treatment avoids structural modifications of the hydrogel by plasma exposure). Alignment and contact of the pressure sensor chip and the PDMS layer to enable covalent bonding.

The process time after plasma treatment can be extended by extracting the un-cross-linked components in the polymer bulk phase using organic solvents (e. g. diisopropylamine) prior to plasma treatment [50]. The bonding strength can be enhanced by an additional annealing step.
This results in a sensor element which can be versatilely applied to a number of different package types and which can be completed with actuator elements to provide the compensation pressure. The sensor element was mounted on a modified TO8 package using an epoxy-based adhesive (Fig. 6.14). Electrical interconnects between the terminals of the WHEATSTONE bridge circuit on the top side of the pressure sensor chip and the pins on the package were established by ultrasonic wire bonding. The top side of the TO8 package was hermetically closed with a custom-build metallic pressure chamber exhibiting a connector port for compressed air where the pneumatic actuator can be connected.

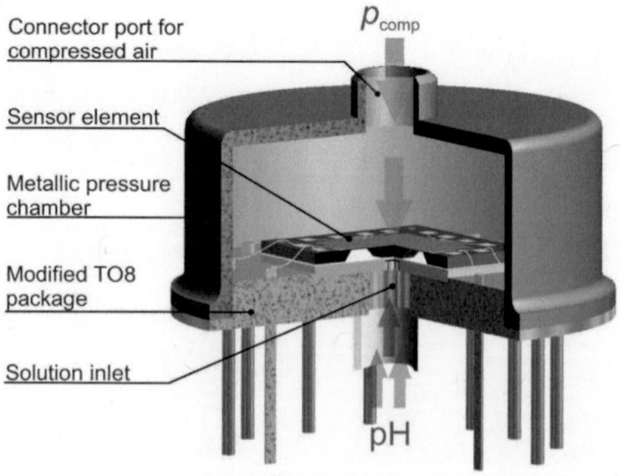

Figure 6.14: Cross-sectional view of the sensor element mounted on a modified TO8 package and hermetically enclosed in a custom-made metallic pressure chamber.

To evaluate the alignment of the center boss with respect to the cavity after assembly, X-ray imaging was used as a non-destructive, reliable method for visualizing the interface between the center boss structure and the cavity (Fig. 6.15).

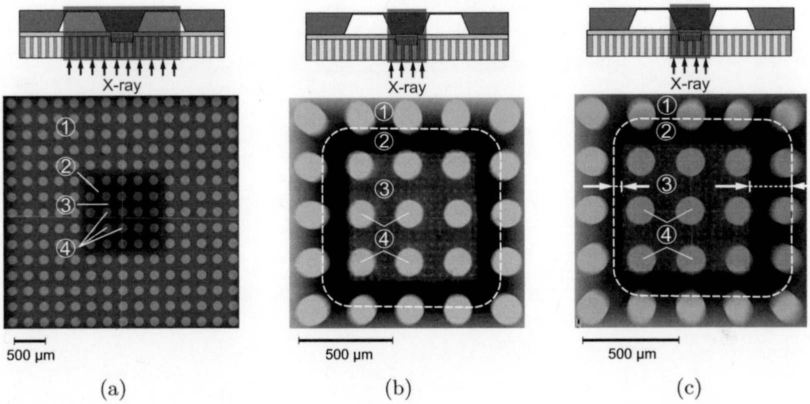

Figure 6.15: X-ray images of the sensor element: (a) sensor backside, (b) enlarged view of the interface between the square cavity and the boss structure from (a) where the boss structure is properly aligned with respect to the square cavity, (c) horizontal misalignment of the boss structure with respect to the square cavity, indicated by white arrows. The absence of an exactly sharp edge of the boss can be attributed to the existence of the sloped flanks of the center boss and the resulting decreasing thickness for X-ray transmission. 1 rigid backplate, 2 bottom face of the boss structure, 3 square cavity containing the hydrogel layer, 4 microhole array.

6.3 Sensor in closed-loop configuration and its components

The actual closed-loop sensor is developed by integrating the sensor element into the feedback structure. In the following, a model structure of the closed-loop system is specified and the single components of the compensation circuit are described and characterized.

6.3.1 Model of the system structure

According to the system-based description of the compensation method in Section 5.3, a model structure in the form of a linear control circuit is assumed for the closed-loop sensor (Fig. 6.16). The sensor element and the measuring unit are considered to be in the forward path and the controller and the actuator are considered as feedback path. The application of the analyte solution with a certain pH value leads to the conversion of a concentration into an osmotic swelling pressure and – if not suppressed – to a displacement w of the gel layer (chemo-mechanical conversion). This displacement is detected by the pressure transducer and converted into an electrical difference voltage

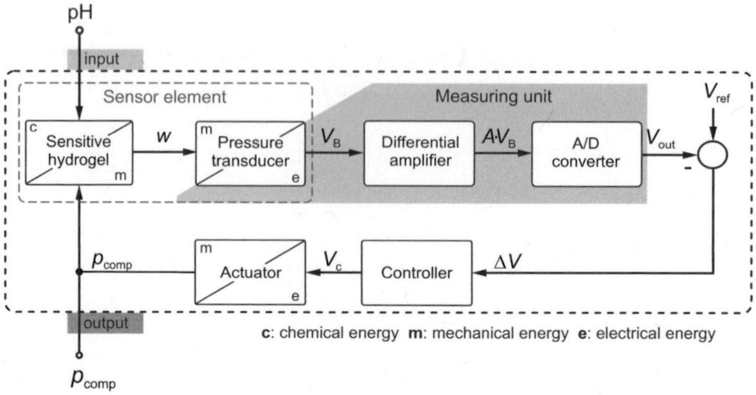

Figure 6.16: Model structure of the compensation circuit of the hydrogel-based pH sensor in closed-loop configuration.

V_B (mechano-electrical conversion). The difference voltage is amplified (amplification factor A) and subsequently converted into a digital signal V_{out}. The comparison with a pre-set reference voltage V_{ref} leads to a difference voltage $\Delta V = V_{ref} - V_{out}$ which is delivered to the controller. The control signal V_c is passed to the actuator providing a pressure p_{comp} (electro-mechanical conversion) which counteracts the displacement of the gel layer and, thus, the swelling pressure.

6.3.2 Measuring unit

The measuring unit is of particular importance for a proper sensor operation, because reliable and sensitive detection even of small displacements w of the hydrogel layer with minimal lag time is necessary. This is primarily accomplished by the pressure transducer which exhibits a sensitivity of about 20 mV/kPa and a non-linearity of less than 5.7 %, respectively, within the linear part of the characteristic curve ($0 < p < 3$ kPa) (Fig. 6.18a). A 12-bit analog digital converter (ADC) (LabJack U3-HV, Lakewood, USA) which can communicate with a PC via USB and comprises 16 flexible I/O ports, 4 additional digital I/O ports and 2 analog outputs was used for data acquisition. Depending on the number of allocated ports, sampling times of 0.6 ms to 4 ms are realized [209]. The bridge voltage V_B of the pressure transducer needs to be continuously measured which is done by the analog input ports of the ADC in the range of $-10\,\mathrm{V} < V_{I,ADC} < 10\,\mathrm{V}$. Thus, the quantization error can be estimated by

$$f_Q = \frac{\Delta V}{2^n} = \frac{V_{I,ADC,max} - V_{I,ADC,min}}{2^n} \tag{6.5}$$

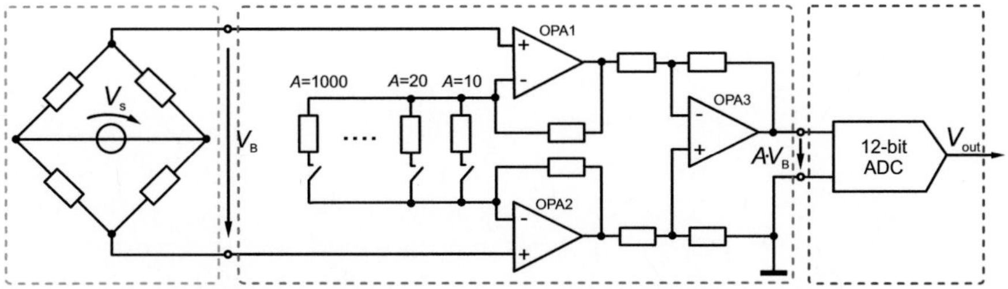

Figure 6.17: Measuring unit comprising piezoresistive WHEATSTONE bridge circuit, DC differential amplifier, and 12-bit analog-to-digital converter. The amplifier circuit is oversimplified, the actual amplifier circuit additionally contains decoupling capacitors and protection diodes.

and amounts to 4.9 mV for $n = 12$ bit. Since typical values of the bridge voltage are in the mV-range and small changes need to be detected, the bridge voltage has to be amplified prior to analog-digital conversion. For this purpose, a direct current precision differential amplifier was developed. The amplifier basically consists of an operational amplifier circuit based on the amplifier module INA114P (Fig. 6.17). The amplification factor A is variable and can be tuned stepwise by a variable resistive input circuit by two orders of magnitude ($A = 10 \ldots 10^3$). The amplifier shows a linear transfer characteristics (Fig. 6.18b). The application of an amplifier reduces the quantization error by factor A to $f'_Q = f_Q/A$ but also reduces the measurement range such that a compromise between an appropriate measurement range and preferably small quantization error has to be attained.

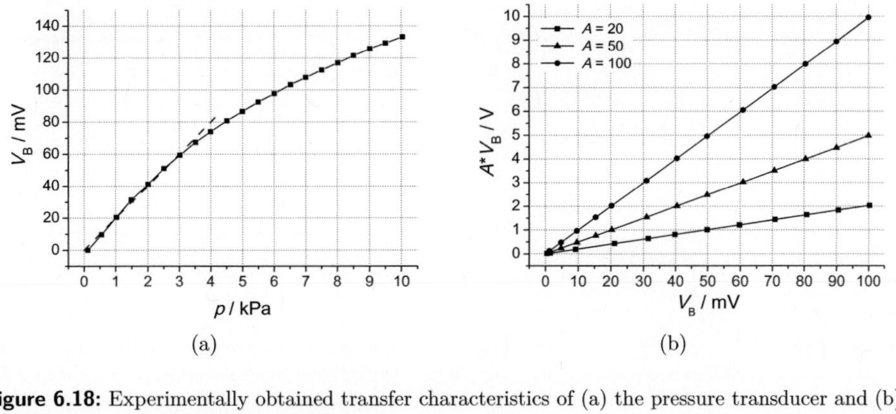

Figure 6.18: Experimentally obtained transfer characteristics of (a) the pressure transducer and (b) the differential amplifier with variable gain.

6.3.3 Controller

For the implementation of the controller, a software solution was favored due to its great flexibility, i. e. variable choice of the controller parameters and easy application of different controller types. The controller was implemented in a LabVIEW environment on a PC. The difference signal which results from subtraction of the output voltage V_{out} from the reference voltage V_{ref} is the controller input and is converted into the control signal V_c according to the control law. Based on the theoretical analysis in Section 5.3, a PI-controller was chosen and implemented to reduce the sensor response time and to suppress linear dynamic drift effects. The communication between the controller and the actuator is accomplished by a RS-232 serial interface. The time necessary to calculate the control signal is much smaller than the time for data transfer between the controller and the actuator. Based on preliminary experiments, a cycle time of 1 s has been assessed which is much faster than the reaction time of the hydrogel layer, such that the controller can be assumed as quasi-continuous.

6.3.4 Actuator

The actuator is assigned to convert the control signal – forwarded by the controller – into a defined pressure signal. The selection of the actuator was initially made under the aspects of simple applicability to the sensor element, small response time, and a linear, reliable, precise pressure supply in a broad pressure range. Therefore, a pneumatic actuator (DPI 510, GE Sensing), which provides a controlled pressure of up to 800 kPa above atmospheric pressure by means of a compressor and controlled

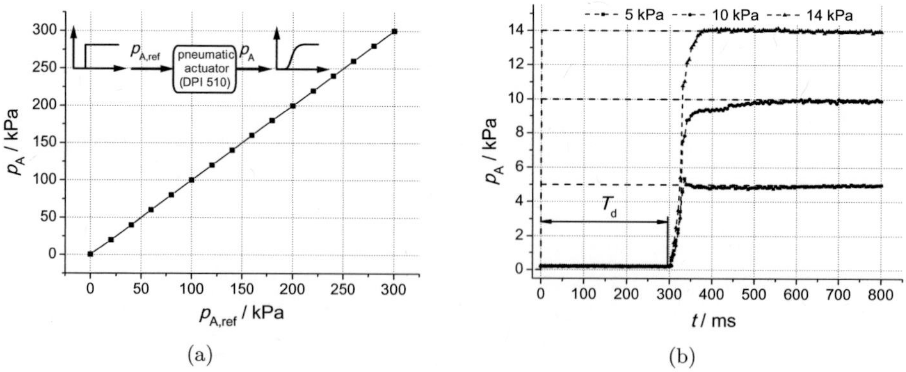

Figure 6.19: Experimentally obtained curves of the precision pneumatic actuator: (a) transfer characteristic in the pressure range between 0 kPa and 300 kPa (the uncertainty bars are smaller than the symbols), (b) step response of selected pressure steps, measured by means of a piezoresistive pressure sensor.

solenoid valves, was used. To verify the transfer characteristics and the dynamics of the actuator, input steps corresponding to different pressures were passed to the actuator by the controller. The transfer characteristics shows linear behavior (Fig. 6.19a). The time-dependent output signal of the actuator was measured by a piezoresistive pressure sensor for different pressure steps (Fig. 6.19b). It is assumed that the dynamics of the pressure sensor is much faster than the actuator because the mechanical inertia of the solenoid valves exceeds that of the flexure plate. Thus, the obtained data truly represents the actuator dynamics. In particular, response times t_{90} below 50 ms have been found for the pneumatic actuator. Moreover, a lag time T_d between the application of the input control signal and the reaction of the actuator has been observed and found to be $T_d = (298 \pm 7)$ ms. This may result from the time needed for data transfer between the controller and the actuator as well as from a mechanical switching delay of the valves.

6.3.5 Hydrogel layer

For PVA/PAA hydrogels, a VPT can be observed for pH changes above the apparent dissociation constant of PVA/PAA, which is close to the dissociation constant of the pH-sensitive functional group ($pK_{a,PAA} \approx 4.7$). Below the $pK_{a,PAA}$, nearly no volume change can be observed because all carboxyl groups exist in their undissociated form. As a result, the free-swelling characteristic curve exhibits a non-linear sigmoid shape with a fairly linear region between pH 5 and pH 9 (Fig. 6.20). Since the occurring protonation-deprotonation reactions are not fully reversible, the acidic-to-basic curve differs with respect to the basic-to-acidic curve, i. e. a distinct hysteretic behavior can be observed [104]. However, restriction of the working range within the linear part of the phase transition region leads to a reduced hysteresis [210]. In this region,

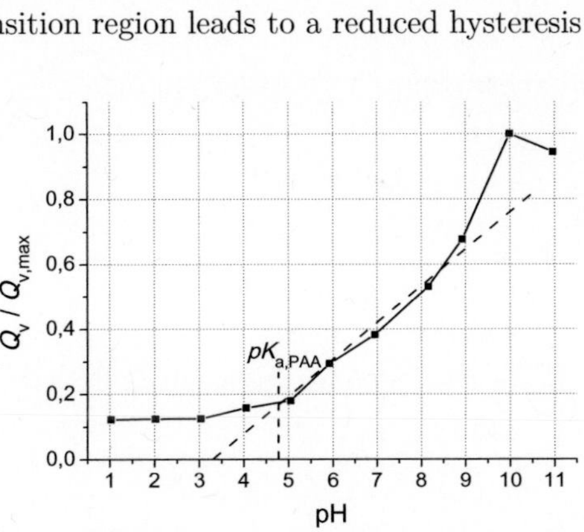

Figure 6.20: Normalized volume swelling degree as a function of pH of a PVA/PAA hydrogel layer under isobar conditions.

the hydrogel material can be assumed to show almost linear behavior.

6.4 System identification and analysis

The model according to Fig. 6.16 will be refined by identification of the individual model components. The identification procedure uses the experimentally obtained characteristics and the *a priori* knowledge of the system components to build simple LTI sub-models. This allows the analysis of the system behavior by simulation and the estimation of a parameter range where the system is robust and BIBO stable.

6.4.1 Identification of model components

a. Measuring unit

The dynamics of the measuring unit is determined by a first-order low-pass filter because a RC-element is connected to the amplifier input for noise suppression. The transfer function can be written as

$$G_{\mathrm{M}}(s) = \frac{A}{1 + sT_{\mathrm{M}}}, \tag{6.6}$$

where A is the variable gain of the amplifier and T_{M} the dominant time constant. Since the measured signals are assumed to have low dynamics and 50 Hz noise should be reliably suppressed, the cut-off frequency was chosen to be 3.4 Hz in circuit design, leading to $T_{\mathrm{M}} = 47\,\mathrm{ms}$.

b. Controller

Because the PI-controller is implemented in software, the controller can be directly described by the ideal linear controller transfer function

$$G_{\mathrm{c}}(s) = k_{\mathrm{P}} + \frac{k_{\mathrm{I}}}{s}. \tag{6.7}$$

c. Actuator

The dynamics of the actuator is primarily determined by the controlled solenoid valve. According to [211], this type of valve can be described by a second-order transfer function between pressure difference and excitation current. By additional consideration of the experimentally observed lag-time, the transfer function can be written as

$$G_{\mathrm{A}}(s) = \frac{K_{\mathrm{A}}}{(1 + sT_{\mathrm{A}})^2} \cdot e^{-s \cdot T_{\mathrm{d}}}, \tag{6.8}$$

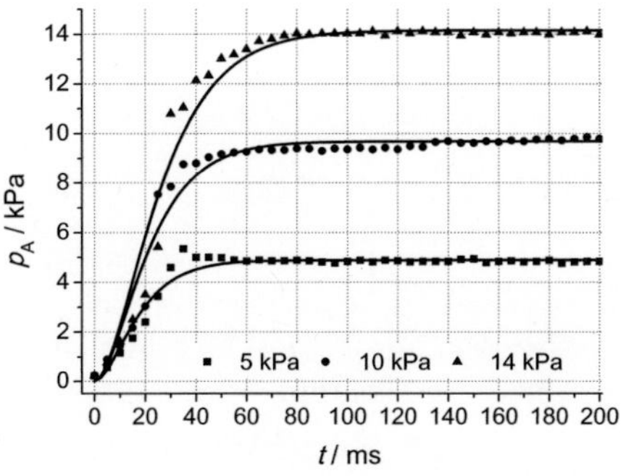

(a)

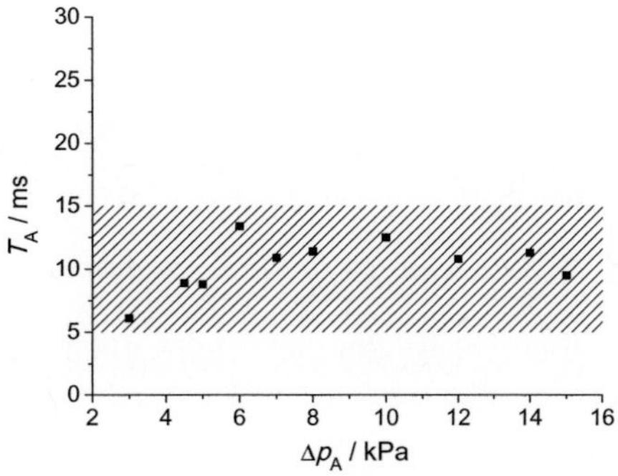

(b)

Figure 6.21: (a) Comparison of the step response of the pneumatic actuator obtained from experimental data (symbols) and the actuator model of Eq. (6.9) (solid lines) (the lag time is not shown). (b) Actuator time constants for different pressure steps estimated from the model.

where K_A is the static gain and T_A and T_d are the time constant and the lag-time, respectively. This leads to a step response in the time domain of

$$p_A(t) = \mathcal{L}^{-1}\left\{G_A(s) \cdot \frac{1}{s}\right\} = K_A \cdot \left[1 - e^{-\frac{t-T_d}{T_A}}\left(1 + \frac{t}{T_A}\right)\right]. \tag{6.9}$$

This equation is in satisfying agreement with the experimental data (Fig. 6.21a). The factor K_A equals the pressure step height Δp_A. The time constant T_A of the actuator slightly differs with respect to the pressure step height (Fig. 6.21b). Thus, considering a safety factor of 2, a general response time $T_A = 25\,\mathrm{ms}$ has been assumed for further considerations.

d. Hydrogel

Physico-chemical processes in hydrogels are complex and, thus, difficult to reduce to a simple model. Therefore, for the hydrogel considered here, a rheological material model was used. The hydrogel is assumed to be composed of two elements describing the thermodynamic behavior and its viscous mechanical properties, respectively. The hydrogel element in Fig. 6.16 can thus be subdivided into two elements (Fig. 6.22). The first element describes the conversion of a pH change into a swelling pressure and the second element describes the macroscopic mechanical response of the gel network. The first element is not influenced by the compensation pressure but only by external concentration changes and is thus located outside the closed-loop structure. Because the counter pressure can be directly act on the gel network structure, the pressure is applied between both elements. Furthermore, the first element is assumed to be

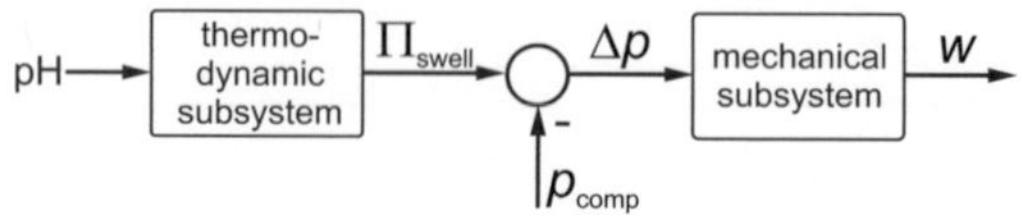

Figure 6.22: Two-block model of the hydrogel describing its thermodynamic and mechanical properties, respectively.

static, i. e. its dynamics is negligible compared to the mechanical response due to cooperative movement of the polymer chains. This simplification is made under the assumption that a pH step instantaneously leads to a swelling pressure.

The mechanical subsystem is modeled on a macroscopic level by means of a rheological gray-box approach. In particular, a four-parameter BURGER-model (cp. Section 2.1.3) is used to describe the experimentally observed phenomena. The corresponding differential equation reads

$$\Delta\ddot{p} + (\frac{E_1}{\eta_1} + \frac{E_1 + E_2}{\eta_2}) \cdot \Delta\dot{p} + \frac{E_1 E_2}{\eta_1 \eta_2}\Delta p = E_1 \cdot \ddot{w} + \frac{E_1 E_2}{\eta_2}\dot{w}. \tag{6.10}$$

Here, Δp is the pressure difference (or stress) applied to the hydrogel due to a pH change, w the deflection of the hydrogel layer, E_1, E_2 and η_1, η_2 are the YOUNG'S moduli and viscosities describing the mechanical behavior of the hydrated polymer chains and the measuring solution, respectively. This model is able to describe the slow cooperative diffusion process as well as the linear dynamic drift effects observed in conjunction with the VPT. Applying the LAPLACE transform to solve the differential equation and subsequent transposition, a second-order transfer function G_s can be derived:

$$G_\mathrm{s}(s) = \frac{w(s)}{\Delta p(s)} = K_\mathrm{s} \cdot \frac{T_1 T_2 s^2 + (T_2 + T_3)s + 1}{T_1 T_2 s^2 + T_1 s} \tag{6.11}$$

with

$$T_1 = \frac{\eta_1}{E_1},\ T_2 = \frac{\eta_2}{E_2},\ T_3 = \frac{E_1 + E_2}{E_1 E_2}\eta_1,\ K_\mathrm{s} = \frac{1}{\eta_1}. \tag{6.12}$$

The step response in the time domain yields

$$w(t) = \mathcal{L}^{-1}\left\{G_\mathrm{s}(s) \cdot \frac{1}{s}\right\} = K_\mathrm{s} \cdot \left[\left(1 - \frac{T_3}{T_1}\right) \cdot \mathrm{e}^{-\frac{t}{T_2}} + \frac{T_3}{T_1} + \frac{t}{T_1}\right]. \tag{6.13}$$

The parameters in Eq. (6.13) are fitted to experimentally observed data for different pH steps (Fig. 6.23). From Fig. 6.23 it can be seen that the slow swelling and shrinking process as well as the linear dynamic drift can be well described. The reciprocal of the time constant T_1 is the drift rate of the linear dynamic drift effect and T_2 is

Table 6.4: Experimentally observed parameter space used for model identification of the hydrogel in the pH range between pH 5 and pH 8.

Parameter	T_1 in s	T_2 in s	T_3 in s
Range	$-8 \cdot 10^4 \ldots 3 \cdot 10^4$	$350 \ldots 1100$	$-2 \cdot 10^4 \ldots 1 \cdot 10^5$

the dominant time constant arising from the slow cooperative swelling/deswelling process. The time constant T_3 cannot be interpreted in such a simple manner. From experiments it was found that the response time and the drift rate of the hydrogel significantly vary depending on the height of the pH step, the pH range, and the direction of the pH change. This was also observed in [104, 109]. Therefore, the real transfer function of the hydrogel $\tilde{G}_s(s)$ differs by an uncertainty $\delta G_u(s)$ from the ideal $G_s(s)$:

$$\tilde{G}_s(s) = G_s(s) + \delta G_u(s). \tag{6.14}$$

For the pH range between pH 5 and pH 8, a parameter space for the parameters in Eq. (6.13) has been derived based on measurements of different pH steps and in different pH ranges (Table 6.4).

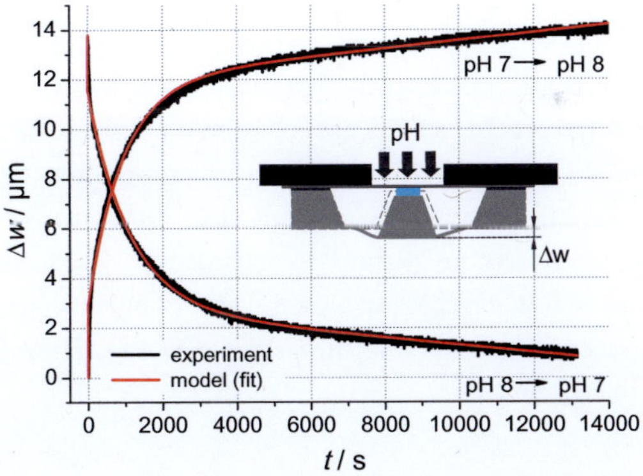

Figure 6.23: Time-dependent change of the hydrogel thickness Δw for application of pH steps between pH 7 and pH 8 and vice versa. The step response arising from the BURGER-model is fitted to the experimental data. The initial hydrogel dimensions (dry state) were $(0.7 \times 0.7 \times 0.05)\,\text{mm}^3$.

e. System model

Based on the identified component models with their experimentally determined parameters, a system model of the closed-loop sensor was implemented in MATLAB/Simulink (Fig. 6.24). It is directly comparable to the system structure in Fig. 6.16. In this model, the measuring unit is split and the quantization noise is

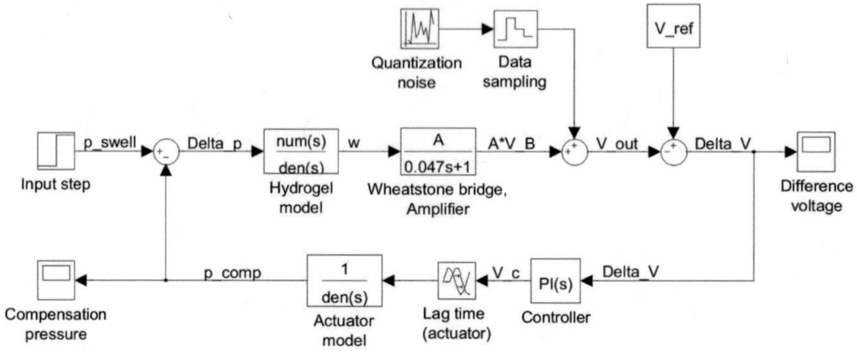

Figure 6.24: Closed-loop system model of the pressure-compensated hydrogel sensor from Fig. 6.16 implemented in MATLAB/Simulink.

modeled by a uniformly distributed random signal together with a zero-order hold element which represents data sampling. The actuator transfer function is modeled by a second-order transfer function in series with a transport delay element.

6.4.2 Stability analysis and parameter estimation

The obtained system model has been used to obtain a controller parameter range were the closed-loop system is stable. Since hydrogel parameters do vary depending on the measurement range, a parameter range, rather than a single set of optimized controller parameters, has been calculated with emphasis on robustness. This means, that the obtained controller parameter range leads to a stable system even if the parameters of the sensor components are subjected to variations. Parameter estimation and stability analysis has been carried out using the "Control and Estimation Tools Manager" which is part of the MATLAB/Simulink package. Here, the following design requirements have been defined:

- Stability (closed-loop),
- Gain margin $g_\mathrm{m} > 7\,\mathrm{dB}$ (open-loop),
- Phase margin $\varphi_\mathrm{m} > 50\,^\circ$ (open-loop),
- Overshoot $< 10\,\%$ (closed-loop step response)

The values for phase and gain margin are suggested in literature to enable a robust system [212]. The limitation of overshoot has been chosen additionally to restrict the sensor output signal in a reasonable way. The parameter range obtained from simulation is shown in Fig. 6.25. Here, two different designs for different phase margins are

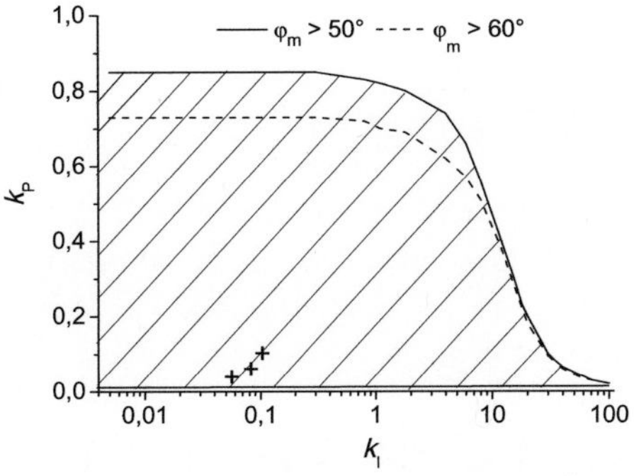

Figure 6.25: Controller parameter range where the system is robust and BIBO stable (ruled region) for different values of the phase margin φ_{m}. The cross marks indicate the set of controller parameters which were applied for the measurement results presented in this work.

shown, revealing that the parameter range is restricted when the robustness requirements are tightened. To enable high robustness, only controller parameters below the dashed line should be applied. The simulation predicts a broad parameter range which can be used for sensor operation and, thus, multiple parameter combinations can be used to obtain the desired sensor behavior. The parameter $k_{\mathrm{P}} = 0$ is excluded for all k_{I} because it would lead to no sensor response. As described in Section 5.3.4 and 5.3.3, in general, an increase of the controller parameters will lead to a decrease in response time and drift rate.

6.5 Measurement procedure and experimental results

6.5.1 General measurement procedure

The set-up used for measurements is shown in Fig. 6.26. Measurements are performed after bringing the sensor and the hydrogel layer, respectively, in a reference state. This includes the following steps:

1. Sealing of the cavity containing the hydrogel: Application of a reference pressure p_{ref} on top of the flexure plate to establish good contact between the center boss and the PDMS intermediate layer;
2. Conditioning: Application of deionized water in order to hydrate the hydrogel such that the cavity is completely filled;
3. Measurement: Application of the pH measuring solution.

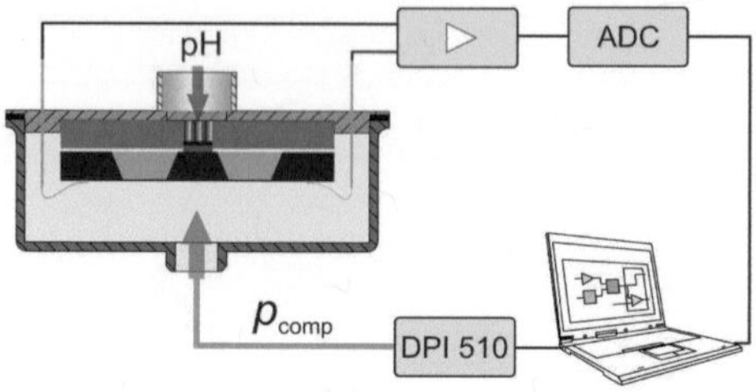

Figure 6.26: Sensor set-up used for measurements.

6.5.2 Reference pressure adjustment

For a proper sensor operation, i. e. the perpetuation of a constant hydrogel volume, a reference pressure p_{ref} – which exceeds ambient pressure – has to be applied to provide a uniform impervious contact between the bottom face of the boss structure and the PDMS layer before the very first contact of an aqueous solution with the hydrogel. In this case, the feedback structure is operated with the condition $p_{\mathrm{swell}} = 0$ and p_{comp} becomes p_{ref}. Since the hydrogel is still in its dry state, the closed-loop performance is independent of the hydrogel behavior ($G_{\mathrm{s}}(s) = 1$) and the step response of the reference pressure yields (cp. Fig. 6.24)

$$p_{\mathrm{ref}}(t) = \mathcal{L}^{-1}\left\{\frac{G_{\mathrm{C}}(s)\cdot G_{\mathrm{A}}(s)}{1+G_{\mathrm{C}}(s)\cdot G_{\mathrm{A}}(s)\cdot G_{\mathrm{IL}}(s)\cdot G_{\mathrm{M}}(s)}\cdot\frac{V_{\mathrm{ref}}}{s}\right\}, \tag{6.15}$$

where $G_{\mathrm{IL}}(s)$ describes the mechanical behavior of the PDMS intermediate layer. The setting of a defined reference voltage V_{ref} proportionally leads to the supply of the reference contact pressure p_{ref} by the actuator to the top of the flexure plate for a constant set of controller parameters (Fig. 6.27). From Fig. 6.27a one can observe that the indicated nominal reference voltage steps are adapted very well by the control circuit and that a stable reference pressure signal is obtained for the observed voltage range (16 mV to 50 mV). The obtained signal do not show any oscillations but monotonic behavior. This can be attributed to the excellent mechanical stability and damping capability of the PDMS intermediate layer. Beside its function as intermediate bonding layer, the PDMS layer also acts as damping element for a reliable control of the reference pressure. In case of a direct contact between the boss and the cavity edge, the boss and the bending plate would act as an undamped mass-spring system which easily leads to oscillations and, hence, is difficult to control. Moreover, the use of a soft but mechanical stable layer results in a sufficiently good sealing of the cavity under constant controlled pressure. From the characteristic curve (Fig. 6.27b), an almost linear dependence between the set reference voltage and the applied reference pressure can be observed. Furthermore,

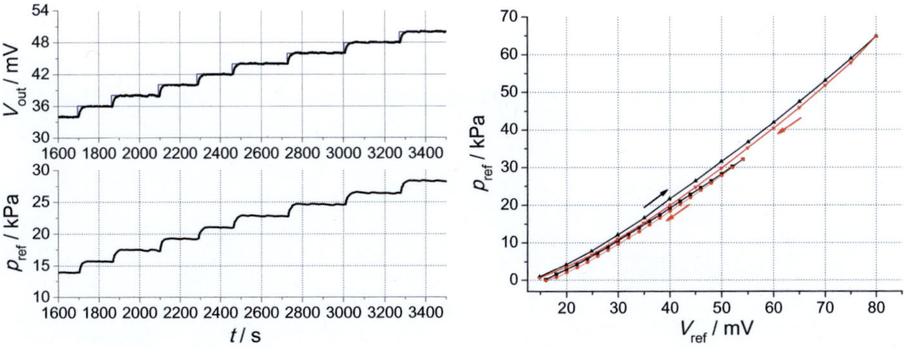

(a) (b)

Figure 6.27: (a) Time-dependent output voltage $V_{\text{out}}(t)$ and reference pressure $p_{\text{ref}}(t)$ for step changes of V_{ref} (blue curve), (b) transfer characteristics $p_{\text{ref}}(V_{\text{ref}})$ ($k_{\text{P}} = 0.06$, $k_{\text{I}} = 0.08$).

subsequent loading and unloading cycles only marginally affect its characteristics. The results suggest that the deformation of the elastomeric PDMS material by compression takes place in the elastic regime for the observed pressure range and no hysteresis due to plastic deformation occurs. The observed deviations are attributed to measurement uncertainties.

To study the influence of an applied reference pressure on the material properties of the PDMS layer for a prolonged time period, different reference voltages and, therefore, different reference pressure values have been initially set and the time-dependent output voltage and reference pressure signal have been measured. From Fig. 6.28 one can observe that the pre-set reference voltages are ideally kept within a narrow range, determined by the quantization noise, for a time period of more than 12 h. The evaluation of the time-dependent pressure signals shows a deviation from the initial reference pressure over time. With regard to the pressure range of the pH measurement and the measurement times (Section 6.5.4), the observed deviations are negligible. That means, that the PDMS material properties only marginally differ with time under compressive load and material effects like creep or relaxation will have no significant influence on the sensor properties in the observed pressure range.

To enable that the desired reference pressure takes effect as fast as possible and that a stable pressure signal is established at the steady state, tuning of the controller parameters is reasonable for optimization. Since the transfer function G_{IL} in Eq. (6.15) is unknown, an experimental optimization approach, i. e. the evaluation of the step response according to Eq. (6.15), is evaluated. Therefore, the response of p_{ref} to a step of V_{ref} have been recorded for different controller parameters. Overshoot and response time τ were determined from the experimental data by fitting the obtained

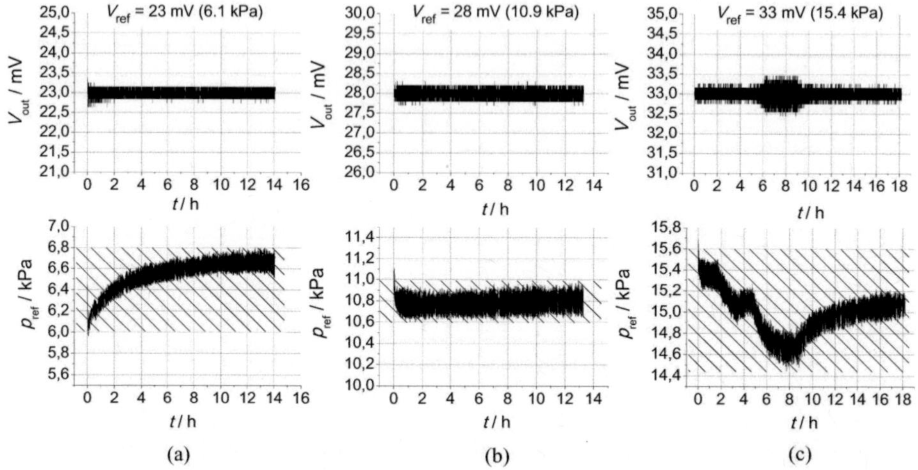

Figure 6.28: Long-term study of the time-dependent output voltage $V_{\text{out}}(t)$ and reference pressure signal $p_{\text{ref}}(t)$ for a pre-set reference voltage V_{ref} of (a) 23 mV, (b) 28 mV, and (c) 33 mV.

curves to the step response of a first-order transfer function. Fig. 6.29 reveals that an increase of k_I for k_P = const. leads to increasing system dynamics and, consequently, to a decrease in response time. However, an increase above $k_\text{I} = 0.6$ causes overshoot (Fig. 6.29a). Further increase of the I-part even leads to oscillations of the step response curve around the desired reference value, which was observed for $k_\text{I} > 1$ (data not shown) and is an undesired system behavior. In contrast, an increase of the P-Part for k_I = const. shows no influence on the response time. Overshoot can be observed for $k_\text{I} > 0.2$ and oscillations have also been found to develop above this

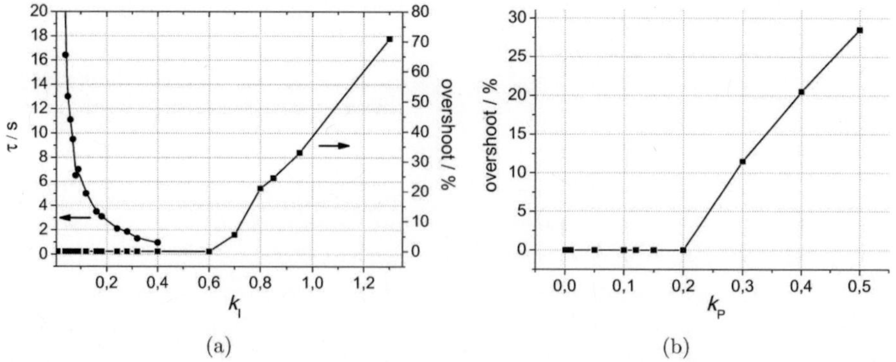

Figure 6.29: Influence of controller parameters on response time τ and overshoot: (a) for the I-part of a PI-controller ($k_\text{P} = 0.06$), (b) for the P-part of a PI-controller ($k_\text{I} = 0.5$).

value (Fig. 6.29b).

In conclusion, for the parameter range $0 < k_P \leq 0.2$ and $0 \leq k_I \leq 0.6$, a stable response without overshoot is obtained for reference tracking. For $k_I > 0.4$, response times of less than 1 s can be achieved. This optimized controller parameter range is specific for the particular sensor set-up described in Section 6.2. These parameters depend on the elastic material properties of the PDMS layer, which can be altered by different monomer-to-crosslinker ratios. Furthermore, the thickness of the flexure plate of the pressure sensor chip as well as the dimensions and geometry of the center boss will have an influence on the absolute values and the range of optimal controller parameters.

6.5.3 Sensor conditioning

After sensor assembly, the cavity is not completely filled by the dry hydrogel layer. In order to fill the whole cavity and, thus, establish a constant hydrogel volume, the hydrogel has to be initially hydrated in deionized water prior to supply of the pH measuring solution. An ideal match of the dry hydrogel volume with the volume of the cavity would not cause any sensor effect (see Section 6.5.4).

The time-dependent sensor response to deionized water for dry hydrogel layers is shown in Fig. 6.30. A reference pressure $p_{\mathrm{ref}} = 10.9\,\mathrm{kPa}$ has been set and the time-dependent sensor output signal $p_{\mathrm{ref}}(t)$ as well as the control deviation, i. e. the voltage difference $\Delta V(t) = V_{\mathrm{ref}}(t) - V_{\mathrm{out}}(t)$, have been measured. Note, that $p_{\mathrm{ref}} = 10.9\,\mathrm{kPa}$ is an empirically chosen value based on the results in Fig. 6.28 and is independent of

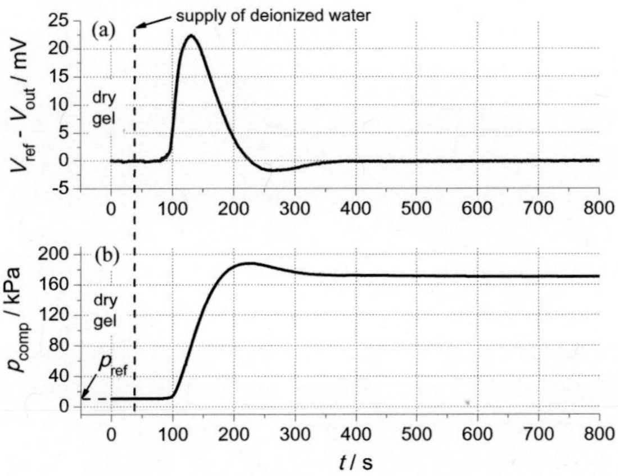

Figure 6.30: Time-dependence of (a) voltage difference $\Delta V = V_{\mathrm{ref}} - V_{\mathrm{out}}$ (control deviation) and (b) sensor output signal p_{comp} for the initial swelling process in deionized water with p_{ref}=10.9 kPa, $A = 50$, $k_P = k_I = 0.1$.

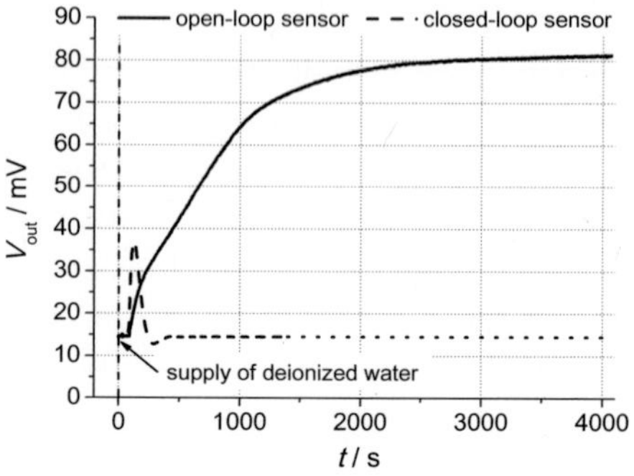

Figure 6.31: Time-dependent electrical output voltage of an open-loop hydrogel microsensor containing a $(0.7 \times 0.7)\,\text{mm}^2$ PVA/PAA hydrogel layer with a thickness of about 50 µm... 60 µm (solid line). For comparison, the time-dependent output voltage change of a closed-loop sensor from Fig. 6.30a is also shown (dashed line).

the applied measuring solution. In principle, also other values of p_{ref} could be used as long as they belong to the linear part of the $p_{ref}(V_{ref})$-curve (cp. Fig. 6.27b). Before application of deionized water, the flexure plate is in its steady state determined by the reference pressure. After about 40 s of deionized water supply, a drop in the difference voltage, which indicates a movement of the flexure plate due to hydrogel swelling, can be observed. During this time span, fluid flow into the microholes and subsequent expansion of the hydrogel takes place. The deflection of the flexure plate is immediately counteracted by an increase in compensation pressure until the pre-set reference voltage is re-established after a time span of about 270 s. Thus, due to the feedback structure, a complete balance of the voltage difference ΔV and, thus, a constant hydrogel volume is achieved. Furthermore, the swelling state of the hydrogel is proportional to the compensation pressure and not to the hydrogel volume anymore (cp. dashed line in Fig. 6.31). As can be seen from Fig. 6.30a, the use of a PI-controller enables an excellent re-establishment of the set reference voltage (no permanent control deviation) and a stable response for the chosen set of controller parameters.

The comparison of the obtained results with the response of an open-loop hydrogel sensor with the same hydrogel material and dimensions allows direct conclusions of the effect of the closed-loop feedback. The time-dependent sensor output voltage of an open-loop sensor for initial hydrogel swelling in deionized water is shown in Fig. 6.31 and selected parameters are summarized in Table 6.5.

- Comparing of Figs. 6.30 and 6.31 reveals a response time t_{90} of 82 s for the pressure signal of the closed-loop sensor, whereas the response time of the open-loop sensor is about 20 times larger and amounts to 1520 s. This behavior

Table 6.5: Comparison of selected parameters of open-loop and closed-loop pH microsensors based on PVA/PAA hydrogels both with initial swelling in deionized water.

Parameter	Open-loop sensor	Closed-loop sensor
Output signal	output voltage V_{out}	pressure p_{comp}
Response time t_{90}	1520 s	82 s
Normalized drift rate δ *	$8 \times 10^{-4}\,\mathrm{min}^{-1}$	0
Overshoot	none	10.9 %
Settling time t_{s} (5 % band)	2000 s	208 s

* normalized to the output signal difference to enable comparability between the voltage and the pressure signal.

strongly suggests that the long exponential swelling phase of the hydrogel due to cooperative diffusion, which dominates the response time of the open-loop sensor, is suppressed by application of the closed-loop feedback.

- The overshoot observed from Fig. 6.30b may be caused by the behavior of the closed-loop transfer function due to the choice of controller parameters and confirms that the sensor behavior is dominated by the feedback path of the compensation circuit. In the open-loop swelling response of a PVA/PAA hydrogel, overshoot has never been observed.
- The temporal drift of the sensor output signal in open-loop configuration for $t > 3000\,\mathrm{s}$ (Fig. 6.31), arising from the repulsion of the weakly charged network chains, can not be observed in the output signal of the closed-loop sensor (Fig. 6.30b).

6.5.4 pH measurement

a. Dynamic and quasi-static behavior

To study the time-dependent sensor output signal for pH measurement, different pH steps have been applied to the sensor (Table. 6.6). An increasing pH value from pH 5 to pH 6 leads to the hydrogel's attempt to expand which can be observed by an increase in the output voltage of the pressure sensor chip. The feedback path reacts with a delay of about 7 s compared to the change in output voltage such that initial macroscopic movement of the polymer chains is possible. This movement is opposed by the compensation pressure and from $t = 126\,\mathrm{s}$ the voltage difference is decreasing until it is completely balanced ($\Delta V = 0$). Here, no control deviation can be observed and the reference state is entirely re-established. The obtained steady state corresponds to a constant hydrogel volume which is the same as before supplying

Table 6.6: Response behavior of a closed-loop and an open-loop pH microsensor, respectively, for different pH measurement ranges.

Measurement range	Closed-loop sensor	Open-loop sensor	Parameters
pH 5 → pH 6	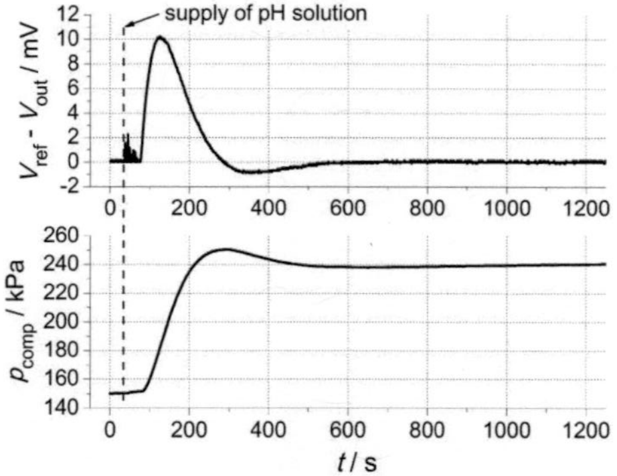	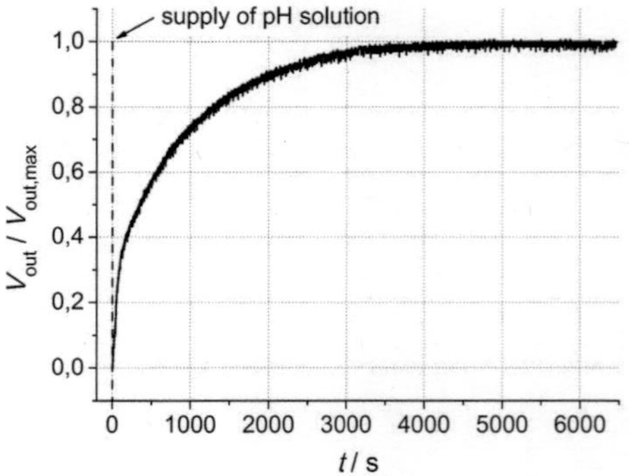	$p_{ref} = 10.9\,kPa$ $A = 50$ $k_P = 0.06$ $k_I = 0.08$
pH 7 ↔ pH 8	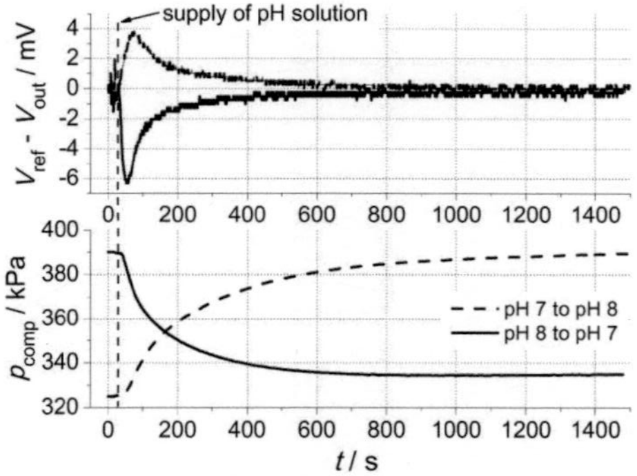	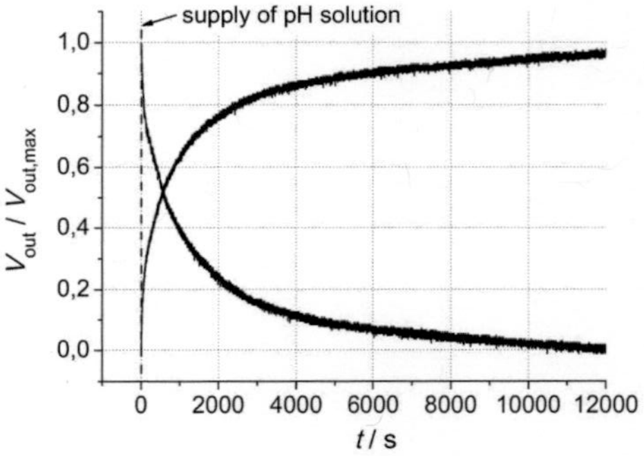	$p_{ref} = 10.9\,kPa$ $A = 20$ $k_P = 0.04$ $k_I = 0.055$

the pH solution and is maintained by a constant compensation pressure of about 240 kPa. As a result, a significant reduction in response time can be observed for the closed-loop sensor because the long exponential swelling phase – as observed for open-loop sensors – does not occur. The applied compensation pressure is a direct measure of the applied pH. The sensor response is in excellent qualitative agreement with the curves obtained for the conditioning process and, thus, clearly confirms an improved sensor performance. A direct comparison of selected performance parameters for open-loop and closed-loop pH microsensors is summarized in Table 6.7.

The obtained results in the acidic pH range are well comparable with results in the basic pH range between pH 7 and pH 8. In this measurement range, the hydrogel has a higher degree of ionization because more carboxyl groups are deprotonated. This leads to an increased repulsion of the charged groups and to a distinct dynamic drift of the sensor signal for an open-loop sensor (Table. 6.6). For the closed-loop sensor, this drift can be clearly decreased. However, even for the closed-loop sensor response, a slight drift can be observed. This result suggests that the hydrogel volume is not ideally constant and hydrogel expansion may be possible to a limited extend. The reason might be that the set of chosen controller parameters is not optimal for this pH range. Different controller parameters might lead to a better drift suppression and an even faster response. The transition from pH 8 to pH 7 leads to protonation of the negatively charged carboxylate groups and consequently to shrinking of the hydrogel layer within the open-loop sensor. Here, also a distinct temporal drift and response times in the order of hours can be observed (Tables 6.6 and 6.7). The corresponding $p_{\mathrm{comp}}(t)$ curve in Table 6.6 shows that the feedback approach works also for decreasing pH values. In this case, the pressure is lowered to re-establish the steady state without the need for a negative pressure. Here, a significant response

Table 6.7: Comparison of selected parameters of open-loop and closed-loop pH microsensors based on PVA/PAA hydrogel.

Sensor type	pH step	Response time t_{90} in s	Normalized drift rate δ in min^{-1}	Overshoot in %	Settling time t_s in s (5 % band)
Closed-loop	pH 5 → pH 6	110	0	11.8	310
	pH 7 → pH 8	680	3.1×10^{-5}	none	980
	pH 8 → pH 7	345	0	none	460
Open-loop	pH 5 → pH 6	2000	0	none	2560
	pH 7 → pH 8	5690	5.8×10^{-4}	none	10360
	pH 8 → pH 7	4360	-6.4×10^{-4}	none	7550

time reduction and a complete drift suppression can be observed in comparison with the output signal of the open-loop sensor. However, the initial starting pressure at pH 7 (324.8 kPa) is not exactly reached again and a deviation of about 10 kPa was observed. This result suggests that the hydrogel volume might not be exactly constant and that swelling into interspaces might occur. These interspaces can be created between the cavity walls and the porous membrane because the membrane dimensions do not exactly match the cavity dimensions due to processing tolerances.

b. Steady-state behavior

Besides the dynamic sensor response, a feature which can be observed from the steady-state signals of the closed-loop sensor is: (i) the signals are in a high pressure range (170 kPa after conditioning in deionized water) and (ii) the relative pressure change after a step change of pH is very large compared to the pressure signal of open-loop sensors [10, 12] which leads to a high sensitivity for closed-loop sensors. This behavior cannot be explained by the structure of the feedback circuit but by the hydrogel behavior under consideration of the FLORY-REHNER theory (Section 2.1.1). Here, the relation between the osmotic swelling pressure Π_{swell} and the volume degree of swelling Q_{v} is described by a set of curves, according to Eq. (2.33), where each curve corresponds to a certain pH value (Fig. 6.32). The curves shown here are not determined experimentally but generated quantitatively based on the FLORY-REHNER equation. The exact shape of the curves varies, depending on the specific hydrogel and the measuring solution. However, the qualitative characteristics generated here is reasonable for interpretation of the measurement results [10, 213]:

- In case of free swelling, the swelling pressure is completely balanced in swelling equilibrium ($\Pi_{\text{swell}} = 0$) and the volume degree of swelling is a measure of the current pH value (gray curve).
- If the hydrogel is confined in an open-loop sensor, the hydrogel cannot expand freely, because free swelling is opposed by mechanical confinement of the package and the flexure plate. Here, the exact path of the VPT depends on the compliance n of the sensor components and the bending plate, respectively (dash-dotted line in Fig. 6.32). Thus, the postulated path for an open-loop sensor would be A $\rightarrow$ B and the resulting pressure signal $\Delta\Pi^{\text{ol}}_{\text{swell}}$ measured by the sensor is the vertical distance between point A and point B.
- For the closed-loop sensor, ideally, the compliance is negligible ($n \rightarrow 0$) because no volume change can occur during a pH transition ($\Delta Q_{\text{v}} = 0$). Thus, the pressure change initiated by a pH change takes place at constant Q_{v}. The postulated path for a closed-loop sensor would thus be C $\rightarrow$ D and the measured

compensation pressure signal equals the pressure difference $\Delta\Pi_{\text{swell}}^{\text{cl}}$ which is the vertical distance between point C and point D.

These considerations reveal that the pressure change of the closed-loop sensor compared to an open-loop sensor is higher and that the measuring signal of the closed-loop sensor can be found in an elevated pressure range. The latter depends on where point C is located. This in turn depends on how close the current swelling state of hydrogel is, compared to its free swelling equilibrium state at the same pH. That directly implies that point C can be varied by the filling degree of the sensor cavity by the dry hydrogel. If the filling degree is low, the hydrated hydrogel is closer to its equilibrium position when the cavity is filled and point C would descend along the $\Pi_{\text{swell}}(Q_{\text{c}})$ curve. If the dry hydrogel would already fill the entire cavity after sensor assembly, no volume swelling could occur ($Q_{\text{v}} = 1$) and point C would ascend towards $\Pi_{\text{swell}} \to \infty$ along the curve. In this case, a pH measurement would not be possible because the different curves (belonging to different pH values) merge at $Q_{\text{v}} = 1$ and, thus, a distinction between different pH values is impossible.

In conclusion, it can be assumed that the pressure range and the absolute compensation pressure signal depend on the hydrogel properties, since they determine the

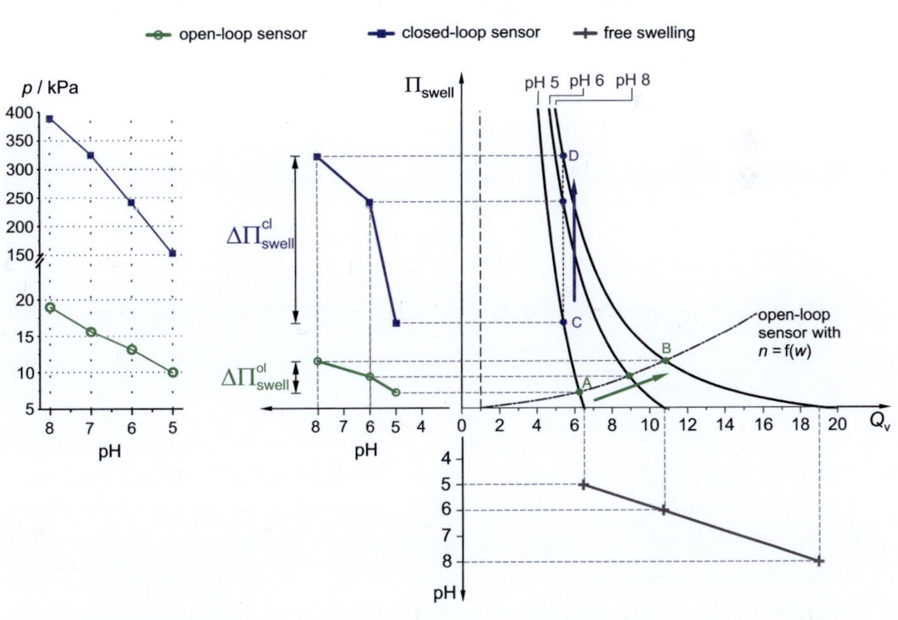

Figure 6.32: Qualitative characteristics of the $\Pi_{\text{well}}(Q_{\text{v}})$-curves according to the FLORY-REHNER theory and postulated paths for pressure generation of open-loop and closed-loop sensors.

exact characteristics of the $\Pi_{swell}(Q_c)$ curves, and the filling degree of the cavity with the dry hydrogel layer. In this way, the sensor properties can be adjusted for specific applications.

6.5.5 Noise suppression

The change of the pH measuring solution from the sensor backside by a pipette causes a random pressure signal to the pressure sensor chip because the former solution is sucked out and the new solution is applied (Fig. 6.33). This leads to disturbances at the sensor input. As can be seen from Fig. 6.33, theses disturbances are propagated in the forward path of the sensor and are directly reflected in a fluctuating change of the signal $V_{out}(t)$ of the pressure sensor chip and, thus, also in the signal ΔV. However, these disturbances are not present in the sensor output signal $p_{comp}(t)$. The feedback circuit with a sufficiently large feedback gain represses input disturbances and leads to a smooth, undisturbed output signal characteristics.

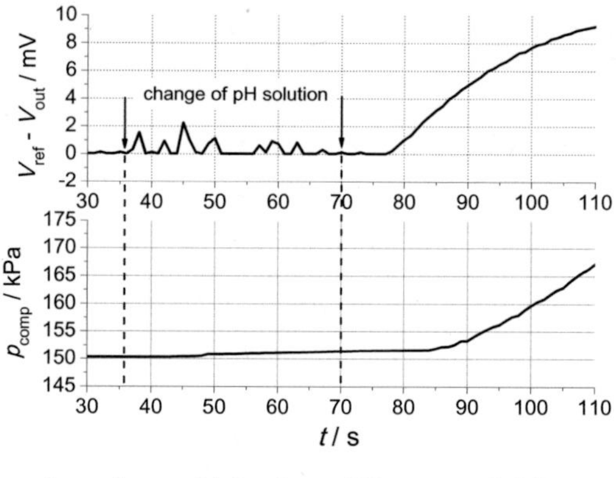

Figure 6.33: Time-dependence of (a) voltage difference and (b) sensor output signal.

6.5.6 Model validation

To validate the quality of the sensor model, the sensor output signal obtained from simulation was compared to the experimentally observed data for different pH steps (Fig. 6.34). For simulation, the same controller parameters as for the measurement have been applied. The model is in good agreement with the experimental data and is able to describe the principle characteristics of the sensor signal. For a pH step from pH 5 to pH 6, the signal $p_{comp}(t)$ is very well described by the simulation, except that the sensor signal reacts slightly delayed and that the settling time is overestimated by the simulation. The simulated signal shows an instantaneous reaction because, due to the application of the BURGER-model, the applied input step is directly forwarded

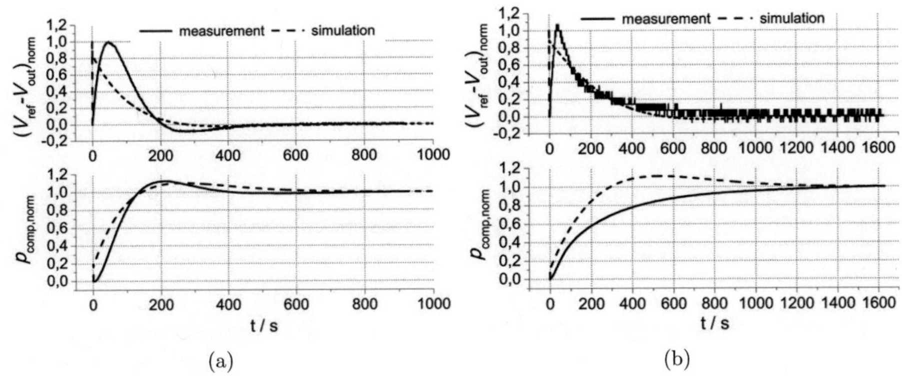

Figure 6.34: Comparison of the measured sensor response with simulation results of the Simulink model for (a) a pH step between pH 5 and pH 6 ($A = 50$, $k_P = 0.06$, $k_I = 0.08$) and (b) a pH step between pH 7 and pH 8 ($A = 20$, $k_P = 0.04$, $k_I = 0.055$).

which can never be the case for the real sensor. However, it is surprising that the simple mechanical model applied for the hydrogel, which neglect all physico-chemical processes inside the hydrogel, is in such good agreement with the measured data. In the real sensor, also diffusion processes of the analyte into or out of the hydrogel layer might occur which are not described by the model structure. Furthermore, the mechanical characteristics of the PDMS intermediate layer is not considered within the model but might superimpose the mechanical behavior of the hydrogel and, thus, influence the closed-loop response. This might explain the undershoot of the time-dependent reference signal $V_{ref} - V_{out}$ in Fig. 6.34a.

For a pH step between pH 7 and pH 8, the deviation between measurement and simulation is more pronounced and the measured signal shows a larger response time compared to the simulation. This behavior strengthens the assumption that the hydrogel layer can still expand in some way inside the cavity and the cooperative movement of the polymer chains cannot be fully suppressed for this measurement.

6.6 Conclusions and summary

This chapter addresses an entirely new approach for hydrogel sensors and describes the design, fabrication, modeling, and test of a closed-loop hydrogel-based pH sensor. Here, a pH-sensitive PVA/PAA hydrogel layer is placed into a square cavity which is sealed by the center boss structure of a silicon pressure sensor chip in order to establish a constant volume. The sensor element is part of a feedback circuit comprising a PI-controller and a pneumatic actuator in the feedback path. In this way, in the steady state, the hydrogel volume is kept constant during the whole measuring process, independent of the applied pH value. The compensation pressure, necessary to maintain this balanced state, is the actual sensor output signal.

The silicon-based technology for fabricating the sensor element allows further miniaturization. A two-step plasma bonding process – using a PDMS intermediate layer – has been found as a reliable way for sensor assembly at low temperatures. Besides its function as bonding layer, the PDMS layer acts as an elastic damping element and enables accurate control of the reference pressure which is applied to seal the cavity.

The sensor components have been modeled individually based on experimental results and a linear model for the closed-loop sensor has been developed out of it and was implemented in MATLAB/Simulink. This model has been used to extract a controller parameter range where the system is robust and stable. The modeling of the hydrogel can be regarded as the crucial step during model development. Here, a simple linear rheological material model, i. e. the BURGER model, has been used to describe the hydrogel behavior. However, more sophisticated hydrogel models might enhance the quality of the sensor model and, thus, the predictability of the experimental results. This could be done using a system identification approach [214] and should be addressed in further studies.

Selected controller parameters from the parameter range obtained by simulation have been applied for pH measurement. A stable sensor operation could be shown in the pH range between pH 5 and pH 8 and the compensation pressure was used as a direct measure of pH. Depending on the respective pH range and the applied controller parameters, a significant response time reduction of about one order of magnitude (factor 8 to 18) could be achieved compared to open-loop sensors. Furthermore, the temporal drift present in the time-dependent output signal of open-loop sensors could be significantly decreased and in some cases completely suppressed. An increased sensitivity in a high pressure range could be demonstrated and it was argued that this behavior depends on the hydrogel itself and can be adapted by the filling factor of the cavity.

IV Summary and future prospects

7 Summary

The present work addresses novel versatile approaches for the improvement of hydrogel-based piezoresistive chemical microsensors which use flexure plate transducers. In particular, the limitations of these sensors like long sensor response times, linear drift of the time-dependent sensor output signal as well as limited mechanical hydrogel stability were intended to be improved. To accomplish this aim, two different approaches have been pursued:

- The application of acrylamide-based hydrogels with improved material properties to fabricate advanced microsensors;
- The application of the force compensation method to hydrogel-based sensors.

7.1 Sensors based on hydrogels with improved material properties

Sensors based on two types of acrylamide-based hydrogels with sensitivity towards different external stimuli have been studied: (i) p(AAm-*co*-AA) hydrogels for pH sensing and (ii) PNIPAAm hydrogels, sensitive towards concentration of organic co-solvents, concentration of salt ions as well as temperature.

For sensors based on p(AAm-*co*-AA) hydrogels, a network of channel-like interconnected macropores in the µm-range has been created inside the hydrogel to increase the surface-to-volume ratio, to enhance the solution transport into the gel, and to decrease the critical hydrogel dimension in order to improve the swelling rate of the gel layer and, thus, the sensor response time. The creation of an interconnected channel structure was successfully established by the porogen leaching technique.

To study the influence of different porogen materials on the swelling rate, four different hydrogel types using four different porogens, namely NaCl, $CaCO_3$, 1,4-dioxane, and PEG with different molecular weight, respectively, have been synthesized. For these gels, the time-dependent mass degree of swelling for step changes of the pH value has been recorded and compared to non-porous hydrogels by free-swelling experiments. Hydrogels synthesized in the presence of PEG with a molecular weight of $6000\,\mathrm{g\,mol^{-1}}$ (PEG6000) showed a significant response time reduction which was most pronounced compared to hydrogels which were fabricated using other porogens. In particular, a weight ratio of 32 wt% PEG was found to be sufficient for the formation of an interconnected pore structure. Therefore, this hydrogel type has been explicitly selected for further characterization and sensor integration.

ATR-FT-IR spectroscopy measurements on these gels revealed that PEG could be completely removed from the network by washing in water for five days. As a result, an open macroporous network structure has been attained which was directly proved by SEM measurements. The SEM results could be directly confirmed by MIP measurements where a narrow pore size distribution with pore diameters between 2 µm . . . 5 µm and an average pore diameter of 3.3 µm were observed.

For the fabrication of hydrogel sensing layers, this hydrogel type has been attached to MPS-modified glass substrates where homogeneous layers of about 70 µm were achieved by filling of a channel with defined height by capillary forces. Single hydrogel-covered glass dies of $(1 \times 1)\,\mathrm{mm}^2$ were produced by wafer sawing and successfully integrated as sensing layers into piezoresistive microsensors. In this way, sensors with good repeatability and reproducibility of the sensor output signal were achieved. The transfer characteristics of sensors with macroporous hydrogel layer was comparable to sensors based on conventional non-porous hydrogels. The absolute sensor output signal was found to be reduced by using macroporous gels due to the reduced polymer volume fraction and the mechanical influence of the flexure plate. The evaluation of the time-dependent sensor output signal revealed a distinct response time reduction of 60 % for increasing and 55 % for decreasing pH values, whereas the smallest response time was found to be 6.5 min (pH 7 to pH 8).

The poor mechanical stability of chemically cross-linked PNIPAAm gel layers, which leads to a reduced long-term stability of the sensor, has been improved by the application of photopolymerizable inorganic Laponite XLS nanoparticles to the organic PNIPAAm matrix. Uniaxial tensile measurements of these hybrid networks revealed very tough materials with extraordinary elongation at break (>1100 %) and high tensile moduli. It was shown that these materials could be successfully transferred to sub-100 µm layers where the formation of a hybrid network structure was proved by EDX and AFM. The latter method revealed a random distribution of nanoparticles in the PNIPAAm matrix. AFM nanoindentation measurements, to study the elastic properties of these thin gel layers, showed that in the hydrated state below the LCST, these materials exhibit very small YOUNG's moduli compared to conventional gels. Above the LCST, elevated values of YOUNG's modulus could be observed, even exceeding that of BIS-gels.

Hydrogel layers with different clay concentrations have been successfully integrated into piezoresistive microsensors. Measurement results revealed that hydrogel layers with low clay concentration (<15 mol%) are not suitable for sensor applications. It was found that the compliance of the flexure plate and the visco-elastic hydrogel properties lead to an almost vanishing sensor signal as well as pronounced drift effects for these materials. However, hydrogels with high clay concentration (NC15, NC20) exhibited very small drift rates and sensor signals with good repeatability. Furthermore, it could be shown by temperature stimulation that the heterogeneous network formation (shrinkage barrier effect) – known from BIS-gels – which leads to very long sensor response times does not occur for NC-gel layers. This was also found for measurement of different ethanol concentrations below 50 vol% and enables a much faster determination of ethanol concentrations compared to sensors based on BIS-gels. For the detection of salt concentrations, in contrast to sensors using BIS-gels, a salting-in effect has been found to determine the sensor response for small NaCl concentrations and salting-out has been found for NaI for the observed concentration range.

7.2 Hydrogel sensors with force compensation

In this work, the force compensation method has been applied to hydrogel-based chemical sensors for the very first time. Essentially, the aim was the perpetuation of a quasi-constant hydrogel volume even if the chemical potential in the surrounding solution is changed. This was intended to suppress the macroscopic cooperative movement of polymer chains in the measuring solution and, thus, decrease the sensor response time as well as reduce long-term drift effects due to chain relaxation.

To get a deeper understanding of the properties and basic structure of sensors operated with the compensation method, this method has been theoretically described by a system-based approach and was directly compared to the properties of sensors operated in deflection mode. In particular, a general sensor model was proposed and three different sensor types, namely PT_1-, PT_2-, and PT_2-D-type sensors have been considered. For these sensor types, the steady-state behavior, the dynamic sensor response (response time, signal characteristics), the influence of systematic static parameter variations as well as the influence of linear drift effects, crucial for hydrogel-based sensors, have been calculated and directly compared for both measurement methods. For the compensation method, the sensor structure was shown to pass into a closed-loop feedback with enhanced system complexity because additional system components, i. e. controller and amplifier, are necessary. Here, three different linear standard controller types, i. e. P, PI, and PID, respectively, have been considered. It was proved that a BIBO-stable system behavior can be achieved for all sensor-controller combinations which is a necessary condition for sensor applications. Additionally, it was shown that a minimum-phase system can be realized for a set of positive controller parameters.

The application of the different sensor-types to a closed-loop feedback structure showed that systematic static parameter variations, like static changes in sensitivity, as well as long-term drift effects can be successfully suppressed. Here, PI-controllers have been found to be an optimal choice for the realization of both tasks. Furthermore, it was shown that a closed-loop compensation circuit, containing a sensor element with characteristic time constant T, can show a significantly reduced time constant if certain parameter constraints are fulfilled. The parameter ranges where a response time reduction can be achieved have been derived for all sensor-controller combinations.

For the implementation of a closed-loop hydrogel-based sensor, a miniaturizable MEMS-based sensor element, containing a pH-sensitive PVA/PAA hydrogel layer, was designed and fabricated. In this design, the hydrogel layer is confined in the square cavity of a silicon backplate which allows solution supply from its backside side and is closed by the center boss structure of a piezoresistive pressure sensor chip from the top side. In this way, a quasi-constant hydrogel volume was achieved. The hydrogel's attempt to change its volume is directly detected by the pressure sensor chip which is part of the measuring unit and enables compact sensor packaging. Reliable bonding between the backplate and the pressure sensor chip was achieved by a two-step plasma bonding process by means of a PDMS intermediate layer. The closed-loop structure was completed with an amplifier with variable gain, a PI-controller, and a pneumatic actuator for supplying the compensation pressure.

The individual components of the compensation circuit have been modeled based on experimentally obtained results and implemented into a general model structure. For the hydrogel layer, a two-block model, consisting of a constant time-independent thermodynamic subsystem and a time-dependent mechanical subsystem, was successfully applied. In particular, the mechanical subsystem was described using a rheological gray-box approach (BURGER-model). The complete system structure was simulated using MATLAB/Simulink with pre-defined stability and robustness specifications, i. e. gain margin, phase margin, overshoot, and an adequate controller parameter range were derived based on the simulation results.

For sensor operation, it was shown that the application of a reference pressure is a mandatory step before the first contact of the device with the pH measuring solution. In this way, the cavity is hermetically sealed such that a constant hydrogel volume is achieved. Here, a set of optimal controller parameters for a fast establishment of the reference state was obtained experimentally. The sensor conditioning process with deionized water to bring the hydrogel from a dry state into a defined reference state provided first evidence that the time constant of the closed-loop sensor is much smaller compared to open-loop sensors. This could be confirmed by pH measurements where a stable sensor response with significantly reduced response time was achieved in the pH range between pH 5 and pH 8. Here, a response time reduction of up to 94 % was observed. Furthermore, the reduction of long-term drift effects could be successfully demonstrated.

8 Future prospects

The results obtained in the previous chapters allow recommendations for additional improvement approaches for the developed sensors.

8.1 Sensors based on hydrogels with improved material properties

For both sensors based on macroporous p(AAm-*co*-AA) hydrogels and nanocomposite PNIPAAm hydrogels, the immobilization of thin hydrogel layers to a substrate should be enhanced to assure reliable, long-term stable layers and, as a result, a long-term stable sensor output signal. Furthermore, this would enable sensors with improved reproducibility and could be achieved by:

- The application of an elastic intermediate layer between the hydrogel layer and the substrate. This layer must not swell in the respective measuring solution but is able to absorb the large bidirectional swelling stress.
- The creation of a cross-linking gradient such that the surface near region of the hydrogel exhibits a higher cross-linking density and, thus, a larger polymer volume fraction at equilibrium. The cross-linking density is then gradually decreasing with increasing distance from the substrate.

Figure 8.1: Microstructured p(AAm-*co*-AA) hydrogel layer on a glass substrate fabricated by microimprinting using a PS stamp.

A successful solution of this challenge would enable additional improvement approaches like structuring of the polymer layers to obtain an increased surface-to-volume ratio and to fabricate small hydrogel structures. Hydrogel patterning can then be done by UV-photolithography, microimprinting (Fig. 8.1) or a combination of both.

Moreover, for sensors based on macroporous hydrogels, the influence of the material porosity on the adhesion properties and the mechanical layer properties should be studied in detail. In particular, the influence of an external force on the pore structure, e. g. the reduction of the pore diameter and, thus, a possible increase in the sensor response time, should be addressed. Based on the results obtained in this work, the synthesis of gels with a defined hierarchical pore structure but a narrow pore size distribution for every pore size level, respectively, is believed to lead to improved material properties. Hence, larger pores would enable a fast solution transport inside the gel along longer distances, whereas the smaller pores would be responsible for a distribution of solvent molecules over short distances. Due to the reduced number of large pores, this approach is believed to improve the mechanical stability and the elastic modulus as well as guarantee a small sensor response time. Such a gel morphology might be achieved by the use of different porogens at the same time.

8.2 Hydrogel sensors with force compensation

The developed closed-loop sensor prototype can be improved with respect to a better qualitative reproducibility of the sensor output signal, a higher sensitivity of the measuring unit, an advanced hydrogel model to obtain improved controller parameters, and an autonomic sensor operation. This can be achieved by:

- A more precise fabrication of the square cavity containing the hydrogel by an anisotropic DRIE process.
- The use of piezoresistive pressure transducers with a block-like boss structure instead of a boss structure with sloped flanks. In this way, the ratio of the stiffened area to the flexible area of the bending plate is reduced, leading to a higher sensitivity of the transducer.
- A general system identification approach, relying on experimentally obtained input-output data (black-box model). Here, e. g. an autoregressive model with exogenous input (ARX-model) could be used to achieve an improved model for the hydrogel [214].
- The development of autonomic, miniaturizable actuators to replace the current pressure source. For that purpose, different actuator principles have been compared with particular respect to their energy density. Two actuator prototypes,

i. e. a bimetal actuator and a phase transition actuator (Fig. 8.2), have been developed and tested but need to be improved further [215].

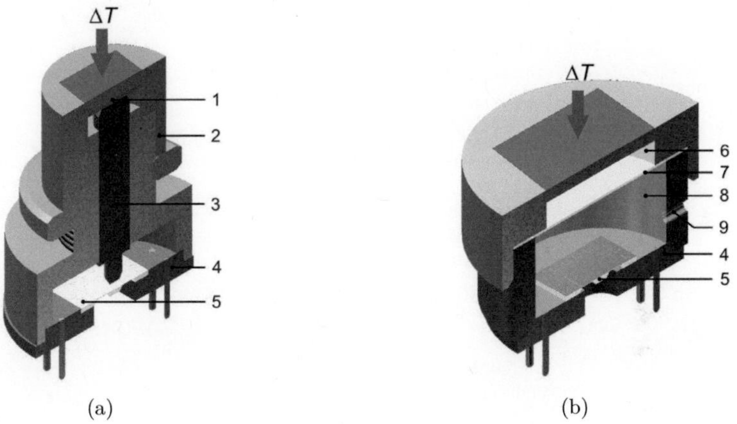

Figure 8.2: Cross-sectional view of the actuator prototypes mounted on a TO8 package with piezoresistive pressure sensor chip: (a) bimetal actuator, (b) phase transition actuator. 1 circular bimetal plate, 2 precision adjustment screw, 3 piston for force transduction, 4 TO8-package, 5 piezoresistive pressure sensor chip with center boss structure, 6 material with liquid-gas phase transition due to temperature excitation (e. g. n-Pentane), 7 flexible membrane, 8 incompressible insulating liquid, 9 inlet [215].

V Appendices

Appendix A: Hydrogel material

A.1 p(AAm-*co*-AA) hydrogels [1]

A.1.1 Synthesis

Conventional non-porous and porous p(AAm-*co*-AA) hydrogels have been synthesized by free radical polymerization of AAm (Fluka Biochemika) and AA (Fluka) with a mole ratio of 70/30. Polymerization was initiated by the redox system NaPS/TEMED (Riedel-de Häen/Sigma-Aldrich) and the polymer chains have been cross-linked using BIS (Merck).

Non-porous hydrogels

For the synthesis of non-porous hydrogels, 1 g AAm and 0.3041 g AA were dissolved in 2 g distilled water and mixed with an aqueous solution of 0.0544 g (TEMED) and 0.0282 g (BIS) in 1.5 g distilled water. The monomer mixture was filled in an ice-cooled three-necked flask. After stirring for 20 min under nitrogen atmosphere, a solution of 0.0865 g NaPS in 0.5 g distilled water was added to the monomer solution.

Porous hydrogels

In principle, the synthesis of porous gels follows the same route as described for non-porous gels, except that the respective porogen P_n (NaCl, $CaCO_3$, 1,4-dioxane, PEG) is added to the aqueous solution of AAm, AA, TEMED, and BIS (Table A.1). The resulting pre-gel solution was as well filled to an ice-cooled three-necked flask and stirred for about 20 min under nitrogen atmosphere before NaPS was added.

The resulting solutions (with and without porogens) were either transferred to cooled cylindrical glass vessels to obtain defined samples for material characterization (free-swelling experiments, SEM, MIP) or to glass slides – modified with MPS – to obtain defined sensing layers.

[1]The hydrogel synthesis was carried out in cooperation with Dr. S. Zschoche and H. P. Zhang, Leibniz Institute of Polymer Research Dresden, Institute for Macromolecular Chemistry.

Table A.1: Ratios of the chemicals used for the synthesis of porous p(AAm-*co*-AA) hydrogels by porogen leaching.

Sample code	AAm in g	AA in g	TEMED in mol% *	BIS in mol% *	NaPS in mol% *	H_2O in g	P_n in wt% *	Note
NaCl-6	1	0.3041	2	1	2	4	6.1	–
NaCl-18	1	0.3041	2	1	2	4	18.4	–
NaCl-24	1	0.3041	2	1	2	4	24.5	–
CaCO3-10	1	0.3041	0.5	1	0.5	4	10	a
CaCO3-100	1	0.3041	0.2	2	0.2	4.14	100	b
Diox-10	0.6156	0.0208	0.2	2	0.2	– [c]	c	–
PEG1-32	1	0.3041	0.5	1	0.5	4	32	–
PEG6-32	1	0.3041	0.5	2	0.5	4	32	d
PEG6-50	1	0.3041	0.5	2	0.5	4	50	–
PEG6-80	1	0.3041	0.5	2	0.5	4	80	–
PEG8-32	1	0.3041	0.5	1	0.5	4	32	–

* Relative to total monomer mass.
[a] Pre-gel solution was neutralized with 2.1 ml NaOH solution (2 M).
[b] Pre-gel solution was neutralized with 2.89 ml NaOH solution (1 M).
[c] Addition of 1.5 ml NaOH (1.9 M) solution and 2.64 ml 1,4-dioxane.
[d] Sample was also synthesized with 1 mol% BIS to study the influence of crosslinker concentration.

A.1.2 SEM characterization

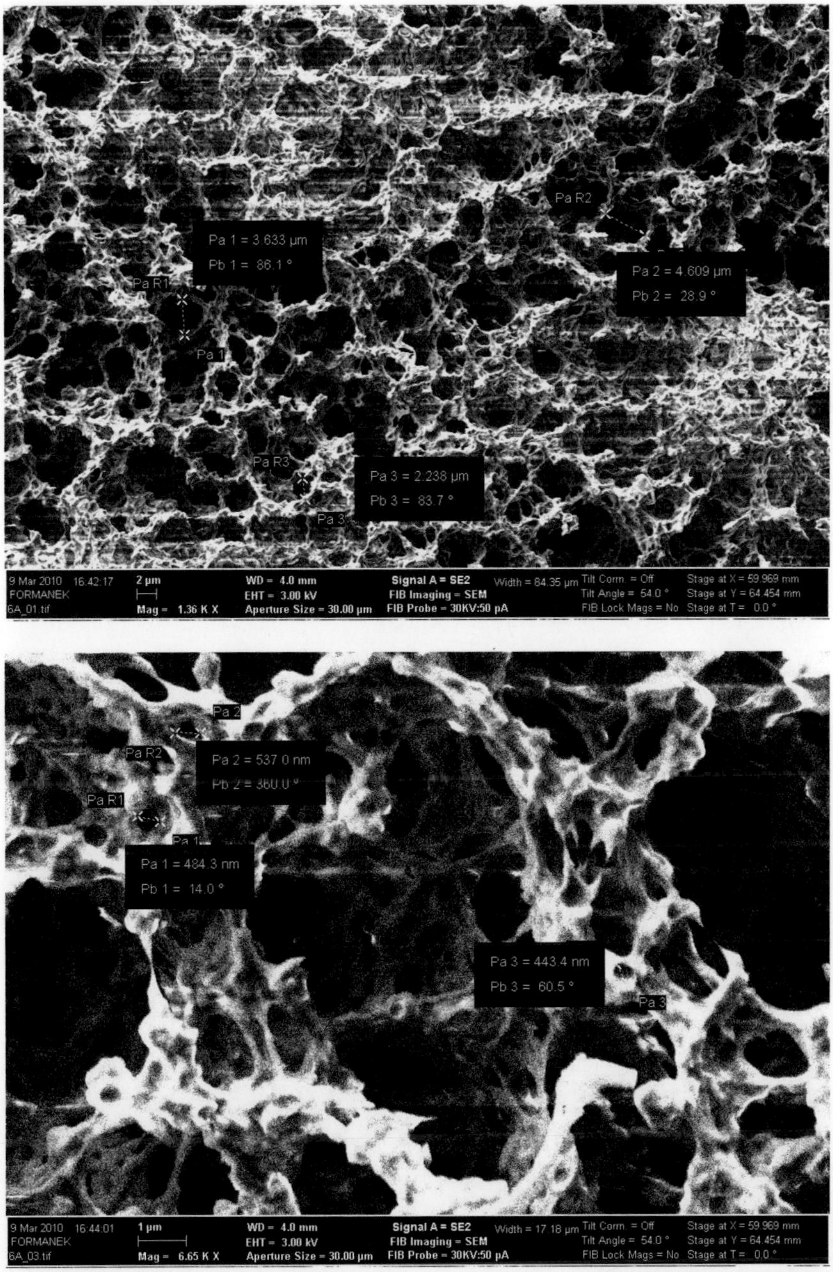

Figure A.1: SEM images of macroporous PEG6-32 hydrogels with different magnifications.

A.2 NC- and BIS-PNIPAAm hydrogels [2]

The monomer NIPAAm, purchased from Acros Co., was purified by recristallization from *n*-hexane and subsequently dried under vacuum. The inorganic clay Laponite XLS consisting of 92.32 wt% $[Mg_{5.34}Li_{0.66}Si_8O_{20}(OH)_4]Na_{0.66}$ + 7.68 wt% $Na_4P_2O_7$ (Rockwood Specialties Inc.) was washed and freeze-dried prior to use. The chemical crosslinker BIS (Fluka) and the photoinitiator 2-hydroxy-4'-(2-hydroxyethoxy)-2-methylpropiophenone (Irgacure 2959) from Aldrich were used as received.

A transparent aqueous solution was prepared by mixing water, crosslinker, NIPAAm, and photoinitiator. The crosslinker was either the inorganic clay (Laponite XLS) for the preparation of NC-gels or BIS for the preparation of chemically cross-linked BIS-gels. For hydrogel synthesis, Milipore water with a conductivity of $18.2\,\mu\Omega/\mathrm{cm}^2$ was used. In order to remove oxygen, the solution was degassed with argon for more than 2 h. The concentrations of photoinitiator and NIPAAm were kept constant with 0.3 wt% photoinitiator and 10 wt% monomer, respectively. The amount of clay was varied to obtain NC6, NC15, and NC20 hydrogels, respectively.
In the notation NCX, X denotes the concentration of clay in mol% with respect to NIPAAm. The concentration X can be calculated as [153, 156]

$$X = \frac{m_{\mathrm{clay}}}{M_{\mathrm{clay}}} \cdot \frac{M_{\mathrm{NIPAAm}}}{m_{\mathrm{NIPAAm}}} \cdot 100, \tag{A.1}$$

where m_{clay}, m_{NIPAAm} are the masses of clay particles and NIPAAm, respectively, and M_{clay}, M_{NIPAAm} are the molar masses of clay particles and NIPAAm, respectively ($M_{\mathrm{clay}} = 747\,\mathrm{g\,mol^{-1}}$, $M_{\mathrm{NIPAAm}} = 113\,\mathrm{g\,mol^{-1}}$).

For the preparation of BIS4- and BIS6-gels, concentrations of 4 mol% and 6 mol% BIS, respectively, with respect to NIPAAm were used for hydrogel synthesis. After dissolving, the solutions, containing clay particles and BIS, were filtered quickly using 0.45 μm nylon-membrane filters.

A.3 PVA/PAA hydrogels [2]

To synthesize PVA/PAA hydrogels, 0.25 g PVA have been added every two hours to 85 g distilled water under stirring at 80 °C until a solution with 15 g PVA was obtained. This corresponds to a solution of 15 wt% PVA. To establish a homogeneous solution, the solution was continuously stirred for three days. PAA was also solved in distilled water, such that a solution with 35 wt% was obtained. The latter was subsequently diluted to a solution with 7.5 wt% PAA. Afterwards, both solutions have been mixed

[2] The hydrogel synthesis was carried out in cooperation with Dr. B. Ferse and A. Grosse, Professorship for Physical Chemistry and Physical Chemistry of Polymers, TU Dresden.

to a polymer solution containing 80 wt% PVA and 20 wt% PAA and stirred for about 48 hours at 45 °C to obtain a homogeneous polymer solution. Within earlier studies of this material system [199], the weight content of the respective polymers in solution and their mixing ratio have been found to be a good compromise between mechanical stability and sensor-actuator functionality of the resulting hydrogels and have been therefore adopted for this work.

Table A.2: Selected properties of the polymers used for the synthesis of PVA/PAA hydrogels as given by the supplier.

Polymer	Supplier	M_w in kg/mol	T_g in °C	Purity in %
Poly(vinyl alcohol) (PVA)	Sigma-Aldrich	89 - 98	51*	99
Poly(acrylic acid) (PAS)	Sigma-Aldrich	450	106 (82*)	99

*determined from static light-scattering experiments in [199]

Appendix B: Supplementary calculations for force compensation

B.1 Application of the ROUTH criterion

To evaluate the sufficient condition for asymptotic stability of a closed-loop transfer function with denominator degree three or higher, the positive coefficients d_ν ($\nu = 0$, 1, 2, ..., n) of the denominator $D_{\text{comp}}(s)$ have to be arranged in the first two lines of the ROUTH table as follows [216]:

$$
\begin{array}{l|ccccc}
\text{n:} & d_n & d_{n-2} & d_{n-4} & \cdots & 0 \\
\text{(n-1):} & d_{n-1} & d_{n-3} & d_{n-5} & \cdots & 0 \\
\hline
\text{(n-2):} & e_{n-1} & e_{n-2} & e_{n-3} & \cdots & 0 \\
\text{(n-3):} & f_{n-1} & f_{n-2} & f_{n-3} & \cdots & 0 \\
. & . & . & . & & \\
. & . & . & . & & \\
. & . & . & . & & \\
3: & g_{n-1} & g_{n-2} & 0 & & \\
2: & h_{n-1} & h_{n-2} & 0 & & \\
1: & i_{n-1} & 0 & & & \\
0: & j_{n-1} & & & &
\end{array}
$$

The coefficients e_{n-1}, e_{n-2}, e_{n-3}, $\ldots$ in the third line result from the cross product of line one and two:

$$
\begin{aligned}
e_{n-1} &= \frac{d_{n-1}d_{n-2} - d_n d_{n-3}}{d_{n-1}}, \\
e_{n-2} &= \frac{d_{n-1}d_{n-4} - d_n d_{n-5}}{d_{n-1}}, \\
e_{n-3} &= \frac{d_{n-1}d_{n-6} - d_n d_{n-7}}{d_{n-1}}, \ \ldots
\end{aligned}
$$

The calculation of the e-values is carried out until all remaining values are zero. The calculation of the f-values is carried out analogous with the two overlying lines as:

$$f_{n-1} = \frac{e_{n-1}d_{n-3} - d_{n-1}e_{n-2}}{e_{n-1}},$$
$$f_{n-2} = \frac{e_{n-1}d_{n-5} - d_{n-1}e_{n-3}}{e_{n-1}},$$
$$f_{n-3} = \frac{e_{n-1}d_{n-7} - d_{n-1}e_{n-4}}{e_{n-1}}, \ \ldots$$

From these two new lines again further lines can be calculated. Note that the last coefficient in the ROUTH table matches the absolute term of the denominator polynomial D_{comp}, i. e. $j_{n-1} = d_0$.

The ROUTH criterion is:

The characteristic polynomial $D_{\text{comp}}(s)$ of the closed-loop transfer function with positive coefficients d_ν ($\nu = 0, 1, 2, \ldots, n$) describes an asymptotic stable system, if and only if all coefficients in the first column of the ROUTH table are positive:
$e_{n-1} > 0, f_{n-1} > 0, \ \ldots \ , g_{n-1} > 0, h_{n-1} > 0, i_{n-1} > 0, j_{n-1} > 0.$

Example: PT$_2$-type sensor with PID-controller

As shown in Table 5.4, the denominator polynomial is $D_{\text{comp}}(s) = T_1^2 s^3 + (T_2 + K^* k_{\text{D}})s^2 + (1 + K^* k_{\text{P}})s + K^* k_{\text{I}}$. Therefore, the complete ROUTH table reads:

3:	T_1^2	$(1 + K^* k_{\text{P}})$	0
2:	$(T_2 + K^* k_{\text{D}})$	$K^* k_{\text{I}}$	0
1:	$\dfrac{(T_2 + K^* k_{\text{D}})(1 + K^* k_{\text{P}}) - T_1^2 K^* k_{\text{I}}}{T_2 + K^* k_{\text{D}}}$	0	
0:	$K^* k_{\text{I}}$		

Thus, the ROUTH criterion is:

$$T_1^2 > 0, \ (T_2 + K^* k_{\text{D}}) > 0, \ K^* k_{\text{I}} > 0, \ (1 + K^* k_{\text{P}}) - \tfrac{T_1^2 K^* k_{\text{I}}}{T_2 + K^* k_{\text{D}}} > 0.$$

The first three requirements are already fulfilled by the necessary condition that all coefficients of the denominator polynomial have to be positive. Hence, the third requirement can be regarded as real sufficient condition that must be additionally considered to guarantee system stability.

B.2 Closed-loop sensor with time-dependent parameter variations

B.2.1 Parameter range for stability and minimum-phase system for drift-afflicted sensors

The parameter range to assure a BIBO-stable and a minimum-phase system for the transfer functions in Table 5.8 in Section 5.3.3 are listed in Table B.1.

Table B.1: Combined parameter range guaranteeing system stability and a minimum-phase system for a drift-afflicted sensor with linear drift.

Controller	Parameter range for a BIBO stable and minimum-phase system
P	$k_\mathrm{P} > 0$
PI	$k_\mathrm{P} \geq 0,\ k_\mathrm{I} > 0$
PID	$k_\mathrm{P} > -\frac{k_\mathrm{I}}{\delta}$ and $k_\mathrm{P} \geq -k_\mathrm{D}\delta,\ k_\mathrm{I} > 0,\ k_\mathrm{D} > 0,\ K(k_\mathrm{P} + \frac{k_\mathrm{I}}{\delta}) > \frac{K^2 k_\mathrm{D} k_\mathrm{I}}{1 + K k_\mathrm{P} + K k_\mathrm{D}\delta}$ [a]

[a]This requirement results from the application of the ROUTH criterion

For the utilization of a PID-controller, the stability requirement arising from the ROUTH criterion can be rewritten as

$$k_\mathrm{P} + K^* k_\mathrm{P}^2 + K^* k_\mathrm{D} k_\mathrm{P} \delta + \frac{k_\mathrm{I}}{\delta} + \frac{K^* k_\mathrm{I} k_\mathrm{P}}{\delta} + K^* k_\mathrm{D} k_\mathrm{I} > K^* k_\mathrm{D} k_\mathrm{I}.$$

From this description it can be easily seen, that the inequation is fulfilled if the controller parameters are positive. Thus, the closed-loop sensor system – comprising a PID controller – is stable, if all controller parameters are positive.

B.2.2 ISE quality criteria for different controller types

Based on the coefficients of the numerator and denominator polynomial of the compensation deviation $\Delta X(s)$, the ISE quality criteria for the different controller types can be calculated according to Table B.2.

Table B.2: Compensation deviation $\Delta X(s)$ as a function of the controller parameters for the three controller types and the rules for the calculation of the ISE performance index J according to [194].

Controller type	$\Delta X(s)$		Performance index J
P	$\frac{1}{(1+K^*k_\mathrm{P})s+K^*k_\mathrm{P}\delta} \mathrel{\hat{=}}$	$\frac{k_0}{l_0+l_1s}$	$\frac{k_0^2}{2l_0l_1}$
PI	$\frac{s}{(1+K^*k_\mathrm{P})s^2+(K^*k_\mathrm{I}+K^*k_\mathrm{P}\delta)s+K^*k_\mathrm{I}\delta} \mathrel{\hat{=}}$	$\frac{k_0+k_1s}{l_0+l_1s+l_2s^2}$	$\frac{k_1^2l_0+k_0^2l_2}{2l_0l_1l_2}$
PID	$\frac{s}{K^*k_\mathrm{D}s^3+(1+K^*k_\mathrm{P}+K^*k_\mathrm{D}\delta)s^2+(K^*k_\mathrm{I}+K^*k_\mathrm{P}\delta)s+K^*k_\mathrm{I}\delta} \mathrel{\hat{=}}$	$\frac{k_0+k_1s}{l_0+l_1s+l_2s^2+l_3s^3}$	$\frac{k_2^2l_0l_1+(k_1^2-2k_0k_2)l_0l_3+k_0^2l_2l_3}{2l_0l_3(l_1l_2-l_0l_3)}$

B.3 Redevelopment of denominator polynomial

As a representative practical example of the method of the redevelopment of the denominator polynomial to verify the possibility of a response time reduction, the combination of a PT_2-sensor with a PI-controller shall be used. This is accomplished using the steps described in Section 5.3.4:

1. Determination of the dominant pole of the open-loop sensor

The dominant pole is

$$s_{\mathrm{p}}^{\mathrm{defl}} = \frac{-T_2 + \sqrt{T_2^2 - 4T_1^2}}{2T_1^2}. \tag{B.1}$$

For the determination of the real part, a case differentiation has to be made:

$$(1)\ \varphi = \frac{T_2}{T_1} \leq 2:\ r_1 = \Re\{s_{\mathrm{p}}^{\mathrm{defl}}\} = -\frac{T_2}{2T_1^2} = \frac{\varphi}{2T_1},$$

$$(2)\ \varphi = \frac{T_2}{T_1} > 2:\ r_2 = \Re\{s_{\mathrm{p}}^{\mathrm{defl}}\} = \frac{-T_2 + \sqrt{T_2^2 - 4T_1^2}}{2T_1^2} = \frac{-\varphi + \sqrt{\varphi^2 - 4}}{2T_1}. \tag{B.2}$$

2. Redevelopment of the denominator polynomial of the closed-loop sensor at point $s = r$

According to Table 5.4, the denominator polynomial of the closed-loop transfer function is

$$D_{\mathrm{comp}}(s) = T_1^2 s^3 + T_2 s^2 + (1 + K^* k_{\mathrm{P}})s + K^* k_{\mathrm{I}}. \tag{B.3}$$

The application of the HORNER scheme to Eq. (B.3) leads to the general representation of the redeveloped polynomial at point $s = r$ as

$$\begin{aligned} D_{\mathrm{comp}}^{\mathrm{r}}(s) = {} & r + K^*(r k_{\mathrm{P}} + k_{\mathrm{I}}) + r^3 T_1^2 + r^2 T_2 \\ & + (2rT_2 + 3r^2 T_1^2 + K^* k_{\mathrm{P}} + 1)(s - r) \\ & + (T_2 + 3rT_1^2)(s - r)^2 \\ & + T_1^2 (s - r)^3. \end{aligned} \tag{B.4}$$

3. Evaluation of the polynomial coefficients

The resulting polynomial coefficients $d_0^{\mathrm{r}}, \dots, d_3^{\mathrm{r}}$ of the redeveloped polynomial are now calculated for the two different cases. Furthermore, the STODOLA condition $d_0^{\mathrm{r}} > 0, \dots, d_3^{\mathrm{r}} > 0$ is applied to calculate the resulting controller parameters.

(1) $r = r_1$:

$$\begin{aligned} d_0^{\mathrm{r}_1} &= \frac{8K^* k_{\mathrm{I}} T_1 - 4\varphi K^* k_{\mathrm{P}} + \varphi^3 - 4\varphi}{8T_1} > 0 \\ d_1^{\mathrm{r}_1} &= \frac{4K^* k_{\mathrm{P}} - \varphi^2 + 4}{4} > 0 \\ d_2^{\mathrm{r}_1} &= -\frac{\varphi T_1}{2} > 0 \\ d_3^{\mathrm{r}_1} &= T_1^2 > 0 \end{aligned} \tag{B.5}$$

From the evalutaion of the parameters it can be easily seen that $d_2^{\mathrm{r}_1} > 0$ can never be fulfilled. Thus, the STODOLA condition is not satisfied. Consequently, a response time reduction is not possible for this case. Hence, further evaluation of the ROUTH criterion is not necessary.

(2) $r = r_2$:

$$\begin{aligned} d_0^{\mathrm{r}_2} &= \frac{K^*(2k_{\mathrm{I}} T_1 + k_{\mathrm{P}}\sqrt{\varphi^2 - 4} - \varphi k_{\mathrm{P}})}{2T_1} > 0 \\ d_1^{\mathrm{r}_2} &= \frac{2K^* k_{\mathrm{P}} - \varphi\sqrt{\varphi^2 - 4} + \varphi^2 - 4}{2} > 0 \\ d_2^{\mathrm{r}_2} &= \frac{T_1(3\sqrt{\varphi^2 - 4} - \varphi)}{2} > 0 \\ d_3^{\mathrm{r}_2} &= T_1^2 > 0 \end{aligned} \tag{B.6}$$

The ROUTH criterion according to Appendix B.1 can be written as

$$d_1^{\mathrm{r}_2} - \frac{d_3^{\mathrm{r}_2} d_0^{\mathrm{r}_2}}{d_2^{\mathrm{r}_2}} > 0, \tag{B.7}$$

and rewritten as

$$\frac{-2K^* k_{\mathrm{I}} T_1 + \sqrt{\varphi^2 - 4}(2K^* k_{\mathrm{P}} + 2\varphi^2 - 6) - 2\varphi^3 + 8\varphi}{3\sqrt{\varphi^2 - 4} - \varphi} > 0. \tag{B.8}$$

The condition $T_1^2 > 0$ is automatically fulfilled. The remaining inequations can now be solved for φ and the respective controller parameters k_{P}, k_{I} as stated in Table 5.10.

Bibliography

[1] PRYPUTNIEWICZ, R.: Current trends and future directions in MEMS. *Experimental Mechanics* 52 (2012), 289–303.

[2] LEO, K.: Organic light-emitting diodes: Efficient and flexible solution. *Nature Photonics* 5 (2011), 716–718.

[3] KUMAR, P.; CHAND, S.: Recent progress and future aspects of organic solar cells. *Progress in Photovoltaics: Research and Applications* 20 (2012), 4, 377–415.

[4] KUCKLING, D.; ARNDT, K.-F.; RICHTER, S.: Synthesis of hydrogels. In: GERLACH, G.; ARNDT, K.-F. (Eds.) *Hydrogel Sensors and Actuators – Engineering and Technology*, chap. 2, Berlin, Heidelberg: Springer (2009) .

[5] GERLACH, G.; ARNDT, K.-F.; SORBER, J.; RICHTER, A.: Sensor zur Messung chemischer Komponenten in Flüssigkeiten. German Patent, DE 101 29 987, 12.06.2001.

[6] GERLACH, G.; GUENTHER, M.; SUCHANECK, G.; SORBER, J.; ARNDT, K.-F.; RICHTER, A.: Application of sensitive hydrogels in chemical and pH sensors. *Macromolecular Symposia* 210 (2004), 1, 403–410.

[7] THONG TRINH, Q.; GERLACH, G.; SORBER, J.; ARNDT, K.-F.: Hydrogel-based piezoresistive pH sensors: Design, simulation and output characteristics. *Sensors and Actuators B: Chemical* 117 (2006), 1, 17–26.

[8] GÜNTHER, M.: *Anwendung polymerer Funktionsschichten in piezoresistiven chemischen und Feuchtesensoren.* Dresden: TUDpress (2009).

[9] GERLACH, G.; ARNDT, K.-F. (Eds.): *Hydrogel Sensors and Actuators: Engineering and Technology.* Berlin, Heidelberg: Springer (2009).

[10] LIN, G.; CHANG, S.; KUO, C.-H.; MAGDA, J.; SOLZBACHER, F.: Free swelling and confined smart hydrogels for applications in chemomechanical sensors for physiological monitoring. *Sensors and Actuators B: Chemical* 136 (2009), 1, 186–195.

[11] Tathireddy, P.; Avula, M.; Lin, G.; Cho, S. H.; Guenther, M.; Schulz, V.; Gerlach, G.; Magda, J. J.; Solzbacher, F.: Smart hydrogel based microsensing platform for continuous glucose monitoring. In: *Conference Proceedings of IEEE Engineering in Medicine and Biology Society*, vol. 2010, 2010, 977–979.

[12] Orthner, M.; Lin, G.; Avula, M.; Buetefisch, S.; Magda, J.; Rieth, L.; Solzbacher, F.: Hydrogel based sensor arrays (2 x 2) with perforated piezoresistive diaphragms for metabolic monitoring (in vitro). *Sensors and Actuators B: Chemical* 145 (2010), 2, 807–816.

[13] Zohuriaan-Mehr, M.; Omidian, H.; Doroudiani, S.; Kabiri, K.: Advances in non-hygienic applications of superabsorbent hydrogel materials. *Journal of Materials Science* 45 (2010), 5711–5735.

[14] Wichterle, O.; Lim, D.: Hydrophilic gels for biological use. *Nature* 185 (1960), 117–118.

[15] Hoare, T. R.; Kohane, D. S.: Hydrogels in drug delivery: Progress and challenges. *Polymer* 49 (2008), 8, 1993–2007.

[16] Lei, Y.; Rahim, M.; Ng, Q.; Segura, T.: Hyaluronic acid and fibrin hydrogels with concentrated DNA/PEI polyplexes for local gene delivery. *Journal of Controlled Release* 153 (2011), 3, 255–261.

[17] Hatefi, A.; Cappello, J.; Ghandehari, H.: Adenoviral gene delivery to solid tumors by recombinant silk-elastinlike protein polymers. *Pharmaceutical Research* 24 (2007), 773–779.

[18] Gustafson, J. A.; Ghandehari, H.: Silk-elastinlike protein polymers for matrix-mediated cancer gene therapy. *Advanced Drug Delivery Reviews* 62 (2010), 15, 1509–1523.

[19] Wheeler, J. C.; Woods, J. A.; Cox, M. J.; Cantrell, R. W.; Watkins, F. H.; Edlich, R. F.: Evolution of hydrogel polymers as contact lenses, surface coatings, dressings, and drug delivery systems. *Journal of long-term Effects of Medical Implants* 6 (1996), 3-4, 207–217.

[20] Boateng, J. S.; Matthews, K. H.; Stevens, H. N.; Eccleston, G. M.: Wound healing dressings and drug delivery systems: A review. *Journal of Pharmaceutical Sciences* 97 (2008), 8.

[21] Atala, A.; Lanza, R. (Eds.): *Methods of Tissue Engineering*. Academia Press (2006).

[22] Van Vlierberghe, S.; Dubruel, P.; Schacht, E.: Biopolymer-based hydrogels as scaffolds for tissue engineering applications: A review. *Biomacromolecules* 12 (2011), 5, 1387–1408.

[23] Kuhn, W.: Reversible Dehnung und Kontraktion bei Änderung der Ionisation eines Netzwerks polyvalenter Fadenmolekülionen. *Cellular and Molecular Life Sciences* 5 (1949), 318–319.

[24] Kuhn, W.; Hargitay, B.; Katchalsky, A.; Eisenberg, H.: Reversible dilation and contraction by changing the state of ionization of high-polymer acid networks. *Nature* 165 (1950), 514–516.

[25] Breitenbach, J. W.; Karlinger, H.: Über Quellung von vernetzter Polymethacrylsäure. *Monatshefte für Chemie / Chemical Monthly* 80 (1949), 312–313.

[26] Katchalsky, A.: Rapid swelling and deswelling of reversible gels of polymeric acids by ionization. *Cellular and Molecular Life Sciences* 5 (1949), 319–320.

[27] Dušek, K.; Patterson, D.: Transition in swollen polymer networks induced by intramolecular condensation. *Journal of Polymer Science Part A-2: Polymer Physics* 6 (1968), 1209–1216.

[28] Tanaka, T.: Collapse of gels and the critical endpoint. *Physical Review Letters* 40 (1978), 820–823.

[29] Okay, O.: General properties of hydrogels. In: Gerlach, G.; Arndt, K.-F. (Eds.) *Hydrogel Sensors and Actuators – Engineering and Technology*, chap. 1, Berlin, Heidelberg: Springer (2009) .

[30] Hulanicki, A.; Glab, S.; Ingman, F.: Chemical sensors definitions and classification. *Pure and Applied Chemistry* 60 (1991), 1247–1250.

[31] Herber, S.; Bomer, J.; Olthuis, W.; Bergveld, P.; van den Berg, A.: A miniaturized carbon dioxide gas sensor based on sensing of pH-sensitive hydrogel swelling with a pressure sensor. *Biomedical Microdevices* 7 (2005), 197–204.

[32] Sheppard, N. F.: Design of a conductimetric microsensor based on reversibly swelling polymer hydrogels. In: *1991 International Conference on Solid-State Sensors and Actuators, TRANSDUCERS '91, Digest of Technical Papers*, 773–776.

[33] Sheppard, N. F.; Tucker, R. C.; Salehi-Had, S.: Design of a conductimetric pH microsensor based on reversibly swelling hydrogels. *Sensors and Actuators B: Chemical* 10 (1993), 2, 73–77.

[34] Sheppard, N. F.; Lesho, M. J.; McNally, P.; Francomacaro, A. S.: Microfabricated conductimetric pH sensor. *Sensors and Actuators B: Chemical* 28 (1995), 2, 95–102.

[35] Lesho, M. J.; Sheppard, N. F.: Adhesion of polymer films to oxidized silicon and its effect on performance of a conductometric pH sensor. *Sensors and Actuators B: Chemical* 37 (1996), 1-2, 61–66.

[36] Rooney, M. T. V.; Seitz, W. R.: An optically sensitive membrane for pH based on swellable polymer microspheres in a hydrogel. *Analytical Communications* 36 (1999), 267–270.

[37] Shakhsher, Z. M.; Odeh, I.; Jabr, S.; Seitz, W. R.: An optical chemical sensor based on swellable dicarboxylate functionalized polymer microspheres for pH copper and calcium determination. *Microchimica Acta* 144 (2004), 147–153.

[38] Oktar, O.; Caglar, P.; Seitz, W. R.: Chemical modulation of thermosensitive poly(N-isopropylacrylamide) microsphere swelling: A new strategy for chemical sensing. *Sensors and Actuators B: Chemical* 104 (2005), 2, 179–185.

[39] Shakhsher, Z. M.; Odeh, I. M.; Rajabi, I. M.; Khatib, M. K.: Optical sensing properties of dithiocarbamate-functionalized microspheres, using a polyvinylpyridine-polyvinylbenzyl chloride copolymer. *Sensors* 10 (2010), 10, 8953–8962.

[40] Endo, T.; Ikeda, R.; Yanagida, Y.; Hatsuzawa, T.: Stimuli-responsive hydrogel-silver nanoparticles composite for development of localized surface plasmon resonance-based optical biosensor. *Analytica Chimica Acta* 611 (2008), 2, 205–211.

[41] Cong, J.; Zhang, X.; Chen, K.; Xu, J.: Fiber optic bragg grating sensor based on hydrogels for measuring salinity. *Sensors and Actuators B: Chemical* 87 (2002), 3, 487–490.

[42] Holtz, J. H.; Asher, S. A.: Polymerized colloidal crystal hydrogel films as intelligent chemical sensing materials. *Nature* 389 (1997), 829–832.

[43] Asher, S. A.; Peteu, S. F.; Reese, C. E.; Lin, M. X.; Finegold, D.: Polymerized crystalline colloidal array chemical-sensing materials for detection

of lead in body fluids. *Analytical and Bioanalytical Chemistry* 373 (2002), 632–638.

[44] LEE, K.; ASHER, S. A.: Photonic crystal chemical sensors: pH and ionic strength. *Journal of the American Chemical Society* 122 (2000), 39, 9534–9537.

[45] ASHER, S. A.; ALEXEEV, V. L.; GOPONENKO, A. V.; SHARMA, A. C.; LEDNEV, I. K.; WILCOX, C. S.; FINEGOLD, D. N.: Photonic crystal carbohydrate sensors: Low ionic strength sugar sensing. *Journal of the American Chemical Society* 125 (2003), 11, 3322–3329.

[46] CHEN, C.; ZHU, Y.; BAO, H.; SHEN, J.; JIANG, H.; PENG, L.; YANG, X.; LI, C.; CHEN, G.: Ethanol-assisted multi-sensitive poly(vinyl alcohol) photonic crystal sensor. *Chemical Communications* 47 (2011), 5530–5532.

[47] WALKER, J. P.; ASHER, S. A.: Acetylcholinesterase-based organophosphate nerve agent sensing photonic crystal. *Analytical Chemistry* 77 (2005), 6, 1596–1600.

[48] WALKER, J. P.; KIMBLE, K. W.; ASHER, S. A.: Photonic crystal sensor for organophosphate nerve agents utilizing the organophosphorus hydrolase enzyme. *Analytical and Bioanalytical Chemistry* 389 (2007), 2115–2124.

[49] ARUNBABU, D.; SANNIGRAHI, A.; JANA, T.: Photonic crystal hydrogel material for the sensing of toxic mercury ions (Hg^{2+}) in water. *Soft Matter* 7 (2011), 2592–2599.

[50] LEE, Y.-J.; BRAUN, P.: Tunable inverse opal hydrogel pH sensors. *Advanced Materials* 15 (2003), 7–8, 563–566.

[51] SHIN, J.; BRAUN, P. V.; LEE, W.: Fast response photonic crystal pH sensor based on templated photo-polymerized hydrogel inverse opal. *Sensors and Actuators B: Chemical* 150 (2010), 1, 183–190.

[52] GRIFFETE, N.; FREDERICH, H.; MAÎTRE, A.; CHEHIMI, M. M.; RAVAINE, S.; MANGENEY, C.: Photonic crystal pH sensor containing a planar defect for fast and enhanced response. *Journal of Materials Chemistry* 21 (2011), 13052–13055.

[53] MILLINGTON, R. B.; MAYES, A. G.; BLYTH, J.; LOWE, C. R.: A holographic sensor for proteases. *Analytical Chemistry* 67 (1995), 23, 4229–4233.

[54] Mayes, A. G.; Blyth, J.; Millington, R. B.; Lowe, C. R.: A holographic sensor based on a rationally designed synthetic polymer. *Journal of Molecular Recognition* 11 (1998), 1-6, 168–174.

[55] Mayes, A. G.; Blyth, J.; Millington, R. B.; Lowe, C. R.: Metal ion-sensitive holographic sensors. *Analytical Chemistry* 74 (2002), 15, 3649–3657.

[56] Marshall, A. J.; Blyth, J.; Davidson, C. A. B.; Lowe, C. R.: pH-sensitive holographic sensors. *Analytical Chemistry* 75 (2003), 17, 4423–4431.

[57] Lee, M.-C.; Kabilan, S.; Hussain, A.; Yang, X.; Blyth, J.; Lowe, C. R.: Glucose-sensitive holographic sensors for monitoring bacterial growth. *Analytical Chemistry* 76 (2004), 19, 5748–5755.

[58] Marshall, A. J.; Young, D. S.; Kabilan, S.; Hussain, A.; Blyth, J.; Lowe, C. R.: Holographic sensors for the determination of ionic strength. *Analytica Chimica Acta* 527 (2004), 1, 13–20.

[59] Marshall, A. J.; Young, D. S.; Blyth, J.; Kabilan, S.; Lowe, C. R.: Metabolite-sensitive holographic biosensors. *Analytical Chemistry* 76 (2004), 5, 1518–1523.

[60] Madrigal González, B.; Christie, G.; Davidson, C. A.; Blyth, J.; Lowe, C. R.: Divalent metal ion-sensitive holographic sensors. *Analytica Chimica Acta* 528 (2005), 2, 219–228.

[61] Bhatta, D.; Christie, G.; Madrigal-González, B.; Blyth, J.; Lowe, C. R.: Holographic sensors for the detection of bacterial spores. *Biosensors and Bioelectronics* 23 (2007), 4, 520–527.

[62] Tan, E. V.; Lowe, C. R.: Holographic enzyme inhibition assays for drug discovery. *Analytical Chemistry* 81 (2009), 18, 7579–7589.

[63] Medlock, K.; Harmer, H.; Worsley, G.; Horgan, A.; Pritchard, J.: pH-sensitive holograms for continuous monitoring in plasma. *Analytical and Bioanalytical Chemistry* 389 (2007), 1533–1539.

[64] Worsley, G. J.; Tourniaire, G. A.; Medlock, K. E. S.; Sartain, F. K.; Harmer, H. E.; Thatcher, M.; Horgan, A. M.; Pritchard, J.: Continuous blood glucose monitoring with a thin-film optical sensor. *Clinical Chemistry* 53 (2007), 10, 1820–1826.

[65] Mao, D.; Dong, L.: Disposable chemical and biological sensors using micro-imprinted stimuli-responsive hydrogel diffraction grating. In: *Solid-State Sensors, Actuators and Microsystems Conference, TRANSDUCERS 2009.*, 152–155.

[66] Ye, G.; Li, X.; Wang, X.: Diffraction grating of hydrogel functionalized with glucose oxidase for glucose detection. *Chemical Communications* 46 (2010), 3872–3874.

[67] Ye, G.; Wang, X.: Glucose sensing through diffraction grating of hydrogel bearing phenylboronic acid groups. *Biosensors and Bioelectronics* 26 (2010), 2, 772–777.

[68] Ye, G.; Wang, X.: Polymer diffraction gratings on stimuli-responsive hydrogel surfaces: Soft-lithographic fabrication and optical sensing properties. *Sensors and Actuators B: Chemical* 147 (2010), 2, 707–713.

[69] Chang, C.; Ding, Z.; Patchigolla, V.; Ziaie, B.; Savran, C.: Reflective diffraction gratings from hydrogels as biochemical sensors. *IEEE Sensors Journal* 12 (2012), 7, 2374–2379.

[70] Tierney, S.; Hjelme, D. R.; Stokke, B. T.: Determination of swelling of responsive gels with nanometer resolution. fiber-optic based platform for hydrogels as signal transducers. *Analytical Chemistry* 80 (2008), 13, 5086–5093.

[71] Tierney, S.; Falch, B. M. H.; Hjelme, D. R.; Stokke, B. T.: Determination of glucose levels using a functionalized hydrogel optical fiber biosensor: Toward continuous monitoring of blood glucose in vivo. *Analytical Chemistry* 81 (2009), 9, 3630–3636.

[72] Tierney, S.; Volden, S.; Stokke, B. T.: Glucose sensors based on a responsive gel incorporated as a fabry-perot cavity on a fiber-optic readout platform. *Biosensors and Bioelectronics* 24 (2009), 7, 2034–2039.

[73] Shakhsher, Z.; Seitz, W. R.; Legg, K. D.: Single fiber-optic pH sensor based on changes in reflection accompanying polymer swelling. *Analytical Chemistry* 66 (1994), 10, 1731–1735.

[74] Zhang, Z.; Shakhsher, Z.; Seitz, W. R.: Aminated polystyrene membranes for a fiber optic pH sensor based on reflectance changes accompanying polymer swelling. *Microchimica Acta* 121 (1995), 41–50.

[75] Zhang, L.; Langmuir, M. E.; Bai, M.; Seitz, W. R.: A sensor for pH based on an optical reflective device coupled to the swelling of an aminated polystyrene membrane. *Talanta* 44 (1997), 9, 1691 – 1698.

[76] Schalkhammer, T.; Lobmaier, C.; Pittner, F.; Leitner, A.; Brunner, H.; Aussenegg, F.: The use of metal-island-coated pH-sensitive swelling polymers for biosensor applications. *Sensors and Actuators B: Chemical* 24 (1995), 1–3, 166–172.

[77] Aussenegg, F.; Brunner, H.; Leitner, A.; Lobmaier, C.; Schalkhammer, T.; Pittner, F.: The metal island coated swelling polymer over mirror system (MICSPOMS): A new principle for measuring ionic strength. *Sensors and Actuators B: Chemical* 29 (1995), 1–3, 204–209.

[78] Seitz, W.: New directions in fibre optic chemical sensors based on polymer swelling. *Journal of Molecular Structure* 292 (1993), 105–114.

[79] McCurley, M.; Seitz, W.: Swelling of polymer membrane for use in a glucose biosensor. *ACS Symposium Series* 487 (1992), 301–309.

[80] Gan, D.; Lyon, L.: Fluorescence nonradiative energy transfer analysis of crosslinker heterogeneity in core-shell hydrogel nanoparticles. *Analytica Chimica Acta* 496 (2003), 1–2, 53 – 63.

[81] Mack, A.; Mao, J.; McShane, M.: Transduction of pH and glucose-sensitive hydrogel swelling through fluorescence resonance energy transfer. In: *Sensors 2005, IEEE*, 912–915.

[82] Mao, J.; McShane, M.: Transduction of volume change in pH-sensitive hydrogels with resonance energy transfer. *Advanced Materials* 18 (2006), 17, 2289–2293.

[83] Kanekiyo, Y.; Sano, M.; Iguchi, R.; Shinkai, S.: Novel nucleotide-responsive hydrogels designed from copolymers of boronic acid and cationic units and their applications as a QCM resonator system to nucleotide sensing. *Journal of Polymer Science Part A: Polymer Chemistry* 38 (2000), 8, 1302–1310.

[84] Richter, A.; Bund, A.; Keller, M.; Arndt, K.-F.: Characterization of a microgravimetric sensor based on pH sensitive hydrogels. *Sensors and Actuators B: Chemical* 99 (2004), 2–3, 579–585.

[85] Chen, H.-M.; Huang, T.-H.; Tsai, R.-M.: A biotin-hydrogel-coated quartz crystal microbalance biosensor and applications in immunoassay and peptide-displaying cell detection. *Analytical Biochemistry* 392 (2009), 1, 1–7.

[86] Cao, Z.; Guo, J.; Fan, X.; Xu, J.; Fan, Z.; Du, B.: Detection of heavy metal ions in aqueous solution by P(MBTVBC-co-VIM)-coated QCM sensor. *Sensors and Actuators B: Chemical* 157 (2011), 1, 34–41.

[87] Ruan, C.; Zeng, K.; Grimes, C. A.: A mass-sensitive pH sensor based on a stimuli-responsive polymer. *Analytica Chimica Acta* 497 (2003), 1–2, 123–131.

[88] Zourob, M.; Ong, K. G.; Zeng, K.; Mouffouk, F.; Grimes, C. A.: A wireless magnetoelastic biosensor for the direct detection of organophosphorus pesticides. *Analyst* 132 (2007), 338–343.

[89] Alvarez, M.; Lechuga, L. M.: Microcantilever-based platforms as biosensing tools. *Analyst* 135 (2010), 827–836.

[90] Zhang, Y.; Ji, H.-F.; Brown, G. M.; Thundat, T.: Detection of CrO_4^{2-} using a hydrogel swelling microcantilever sensor. *Analytical Chemistry* 75 (2003), 18, 4773–4777.

[91] Liu, K.; Ji, H.-F.: Detection of Pb^{2+} using a hydrogel swelling microcantilever sensor. *Analytical Sciences* 20 (2004), 1, 9–11.

[92] Ji, H.-F.; Yan, X.; McShane, M. J.: Experimental and theoretical aspects of glucose measurement using a microcantilever modified by enzyme-containing polyacrylamide. *Diabetes Technology & Therapeutics* 7 (2005), 6, 986–995.

[93] Peng, R.-P.; Chen, B.; Ji, H.-F.; Wu, L.-Z.; Tung, C.-H.: Highly sensitive and selective detection of beryllium ions using a microcantilever modified with benzo-9-crown-3 doped hydrogel. *Analyst* 137 (2012), 1220–1224.

[94] Zhang, Y.; Ji, H.-F.; Snow, D.; Sterling, R.; Brown, G. M.: A pH sensor based on a microcantilever coated with intelligent hydrogel. *Instrumentation Science & Technology* 32 (2004), 4, 361–369.

[95] Hilt, J. Z.; Gupta, A. K.; Bashir, R.; Peppas, N. A.: Ultrasensitive bioMEMS sensors based on microcantilevers patterned with environmentally responsive hydrogels. *Biomedical Microdevices* 5 (2003), 177–184.

[96] VanBlarcom, D.; Peppas, N.: Microcantilever sensing arrays from biodegradable, pH-responsive hydrogels. *Biomedical Microdevices* 13 (2011), 829–836.

[97] Gonska, J.; Schelling, C.; Urban, G.: Application of hydrogel-coated microcantilevers as sensing elements for pH. *Journal of Micromechanics and Microengineering* 19 (2009), 12, 127002.

[98] STRONG, Z. A.; WANG, A. W.; MCCONAGHY, C. F.: Hydrogel-actuated capacitive transducer for wireless biosensors. *Biomedical Microdevices* 4 (2002), 97–103.

[99] LEI, M.; BALDI, A.; NUXOLL, E.; SIEGEL, R. A.; ZIAIE, B.: A hydrogel-based implantable micromachined transponder for wireless glucose measurement. *Diabetes Technology & Therapeutics* 8 (2006), 112–122.

[100] LEI, M.; BALDI, A.; NUXOLL, E.; SIEGEL, R.; ZIAIE, B.: Hydrogel-based microsensors for wireless chemical monitoring. *Biomedical Microdevices* 11 (2009), 529–538.

[101] SIEGEL, R. A.; YUANDONG, G.; LEI, M.; BALDI, A.; NUXOLL, E. E.; ZIAIE, B.: Hard and soft micro- and nanofabrication: An integrated approach to hydrogel-based biosensing and drug delivery. *Journal of Controlled Release* 141 (2010), 3, 303–313.

[102] GERLACH, G.; GUENTHER, M.; SORBER, J.; SUCHANECK, G.; ARNDT, K.-F.; RICHTER, A.: Chemical and pH sensors based on the swelling behavior of hydrogels. *Sensors and Actuators B: Chemical* 111–112 (2005), 0, 555–561.

[103] GUENTHER, M.; KUCKLING, D.; CORTEN, C.; GERLACH, G.; SORBER, J.; SUCHANECK, G.; ARNDT, K.-F.: Chemical sensors based on multiresponsive block copolymer hydrogels. *Sensors and Actuators B: Chemical* 126 (2007), 1, 97–106.

[104] GUENTHER, M.; GERLACH, G.; WALLMERSPERGER, T.: Non-linear effects in hydrogel-based chemical sensors: Experiment and modeling. *Journal of Intelligent Material Systems and Structures* 20 (2009), 8, 949–961.

[105] HAN, I. S.; HAN, M.-H.; KIM, J.; LEW, S.; LEE, Y. J.; HORKAY, F.; MAGDA, J. J.: Constant–volume hydrogel osmometer: A new device concept for miniature biosensors. *Biomacromolecules* 3 (2002), 6, 1271–1275.

[106] ORTHNER, M.; BUETEFISCH, S.; MAGDA, J.; RIETH, L.; SOLZBACHER, F.: Development, fabrication, and characterization of hydrogel based piezoresistive pressure sensors with perforated diaphragms. *Sensors and Actuators A: Physical* 161 (2010), 1–2, 29–38.

[107] HERBER, S.; EIJKEL, J.; OLTHUIS, W.; BERGVELD, P.; VAN DEN BERG, A.: Study of chemically induced pressure generation of hydrogels under isochoric conditions using a microfabricated device. *The Journal of Chemical Physics* 121 (2004), 6, 2746–2751.

[108] GUENTHER, M.; GERLACH, G.; SORBER, J.; SUCHANECK, G.; ARNDT, K.-F.; RICHTER, A.: pH sensors based on polyelectrolytic hydrogels. In: BAR-COHEN, Y. (Ed.) *Smart Structures and Materials: Electroactive Polymer Actuators and Devices*, vol. 5759, 2005, 540–548.

[109] THONG TRINH, Q.: *Hydrogel-based Piezoresistive pH Sensors*. Dresden: TUD-press (2006).

[110] FLORY, P. J.: *Principles of Polymer Chemistry*. Ithaca, New York: Cornell University Press (1953).

[111] ARNDT, K.-F.; KRAHL, F.; RICHTER, S.; STEINER, G.: Swelling-related processes in hydrogels. In: GERLACH, G.; ARNDT, K.-F. (Eds.) *Hydrogel Sensors and Actuators – Engineering and Technology*, chap. 3, Berlin, Heidelberg: Springer (2009) .

[112] FLORY, P. J.: Statistical thermodynamics of random networks. *Proceedings of the Royal Society A* 351 (1976), 1666, 351–380.

[113] JAMES, H. M.; GUTH, E.: Theory of the elastic theory of rubber. *Journal of the Chemical Society* 1 (1943), 455–481.

[114] DONNAN, F. G.; GUGGENHEIM, E. A.: Exact thermodynamics of membrane equilibrium. *Zeitschrift für Physikalische Chemie A* 162 (1932), 4–5, 346–360.

[115] BRANNON-PEPPAS, L.; PEPPAS, N. A.: Equilibrium swelling behavior of pH-sensitive hydrogels. *Chemical Engineering Science* 46 (1991), 3, 715–722.

[116] GÜNTHER, M.; GERLACH, G.: Hydrogels for chemical sensors. In: GERLACH, G.; ARNDT, K.-F. (Eds.) *Hydrogel Sensors and Actuators – Engineering and Technology*, chap. 5, Berlin, Heidelberg: Springer (2009) .

[117] EICHENBAUM, G. M.; KISER, P. F.; SIMON, S. A.; NEEDHAM, D.: pH and ion-triggered volume response of anionic hydrogel microspheres. *Macromolecules* 31 (1998), 15, 5084–5093.

[118] ARNDT, K.-F.; KNOERGEN, M.; RICHTER, S.; SCHMIDT, T.: *NMR Imaging: Monitoring of Swelling of Environmental Sensitive Hydrogels*. Dodrecht: Springer (2006).

[119] GEHRKE, S. H.: Synthesis, equilibrium swelling, kinetics, permeability and applications of environmentally responsive gels. *Advances in Polymer Science* 110 (1993), 81–144.

[120] LESHO, M. J.; SHEPPARD, N. F.: A method for studying swelling kinetics based on measurement of electrical conductivity. *Polymer Gels and Networks* 5 (1998), 6, 503–523.

[121] TANAKA, T.; HOCKER, L. O.; BENEDEK, G. B.: Spectrum of light scattered from a viscoelastic gel. *Journal of Chemical Physics* 59 (1973), 9, 5151–5159.

[122] LI, Y.; TANAKA, T.: Kinetics of swelling and shrinking of gels. *Journal of Chemical Physics* 92 (1990), 2, 1365–1371.

[123] ECKERT, F.: *Bestimmung der kooperativen Diffusionskoeffizienten von Poly(acrylsäure)-Netzwerken.* Diploma thesis, TU Dresden (2003).

[124] BARNES, H. A.; HUTTON, J. F.; WALTERS, K.: *An Introduction to Rheology.* Oxford: Elsevier (1998).

[125] GERLACH, G.; DÖTZEL, W.: *Introduction to Microsystem Technology – A Guide for Students.* Chichester: John Wiley & Sons (2008).

[126] BECKER, W.; GROSS, D.: *Mechanik elastischer Körper und Strukturen.* Berlin, Heidelberg, New York: Springer (2002).

[127] BAO, M.: *Analysis and Design Principles of MEMS Devices.* Amsterdam: Elsevier (2005).

[128] SCHOMBURG, W. K.: *Introduction to Microsystem Design.* Berlin, Heidelberg: Springer (2011).

[129] ZHANG, H. P.: *Poröse pH-sensitive Hydrogele für piezoresistive Sensoren.* Diploma thesis, TU Dresden (2010).

[130] SCHULZ, V.; ZSCHOCHE, S.; ZHANG, H.; VOIT, B.; GERLACH, G.: Macroporous smart hydrogels for fast-responsive piezoresistive chemical microsensors. *Procedia Engineering* 25 (2011), 1141–1144.

[131] IUPAC: *Compendium of Chemical Terminology.* 2nd edn., Oxford: Blackwell Scientific Publications (1997).

[132] MATTIASSON, B.; KUMAR, A.; GALAEV, I. Y. (Eds.): *Macroporous Polymers: Production Properties and Biotechnological/Biomedical Applications.* Boca Raton: CRC Press (2010).

[133] OKAY, O.: Macroporous copolymer networks. *Progress in Polymer Science* 25 (2000), 6, 711–779.

[134] ANNABI, N.; NICHOL, J. W.; ZHONG, X.; JI, C.; KOSHY, S.; KHADEMHOSSEINI, A.; DEHGHANI, F.: Controlling the porosity and microarchitecture of hydrogels for tissue engineering. *Tissue Engineering: Part B* 16 (2009), 4, 371–383.

[135] CHEN, J.; PARK, H.; PARK, K.: Synthesis of superporous hydrogels: Hydrogels with fast swelling and superabsorbent properties. *Journal of Biomedical Materials Research* 44 (1999), 1, 53–62.

[136] HARRIS, J. M.: Introduction to biotechnical and biomedical applications of poly(ethylene glycol). In: HARRIS, J. M. (Ed.) *Poly(ethylene glycol) Chemistry: Biotechnical and Biomedical Applications*, chap. 1, New York: Plenum Press (1992) .

[137] COATES, J.: Interpretation of infrared spectra, a practical approach. In: MEYERS, R. A. (Ed.) *Encyclopedia of Analytical Chemistry*, Chichester: John Wiley & Sons (2000) 10815–10837.

[138] ZHANG, J.; HAN, F.; WEI, X.; SHUI, L.; GONG, H.; ZHANG, P.: Spectral studies of hydrogen bonding and interaction in the absorption processes of sulfur dioxide in poly(ethylene glycol) 400 + water binary system. *Industrial & Engineering Chemistry Research* 49 (2010), 5, 2025–2030.

[139] WASHBURN, E. W.: The dynamics of capillary flow. *Physical Review* 17 (1921), 273–283.

[140] BLACHER, S.; MAQUET, V.; PIRARD, R.; PIRARD, J.-P.; JÉRÔME, R.: Image analysis, impedance spectroscopy and mercury porosimetry characterisation of freeze-drying porous materials. *Colloids and Surfaces A: Physicochemical and Engineering Aspects* 187-188 (2001), 375–383.

[141] EGGER, C. C.; DU FRESNE, C.; RAMAN, V. I.; SCHÄDLER, V.; FRECHEN, T.; ROTH, S. V.; MÜLLER-BUSCHBAUM, P.: Characterization of highly porous polymeric materials with pore diameters larger than 100 nm by mercury porosimetry and X-ray scattering methods. *Langmuir* 24 (2008), 11, 5877–5887.

[142] HO, S. T.; HUTMACHER, D. W.: A comparison of micro CT with other techniques used in the characterization of scaffolds. *Biomaterials* 27 (2006), 8, 1362–1376.

[143] ROUQUEROL, J.; BARON, G.; DENOYEL, R.; GIESCHE, H.; GROEN, J.; KLOBES, P.; LEVITZ, P.; NEIMARK, A. V.; RIGBY, S.; SKUDAS, R.; SING, K.; THOMMES,

M.; Unger, K.: Liquid intrusion and alternative methods for the characterization of macroporous materials (IUPAC technical report). *Pure and Applied Chemistry* 84 (2012), 1, 107–136.

[144] Giesche, H.: Mercury porosimetry: A general (practical) overview. *Particle & Particle Systems Characterization* 23 (2006), 1, 9–19.

[145] Gale, G. W.; Small, R. J.; Reinhardt, K. A.: Aqueous cleaning surface conditioning processes. In: Reinhardt, K. A.; Kern, W. (Eds.) *Handbook of Silicon Wafer Cleaning Technology*, 2nd edn., chap. 4, Norwich, NY: William Andrew Publishing (2008) .

[146] Basu, S. K.; McCormick, A. V.; Scriven, L. E.: Stress generation by solvent absorption and wrinkling of a cross-linked coating atop a viscous or elastic base. *Langmuir* 22 (2006), 13, 5916–5924.

[147] Hoa, P. L. P.: *Uncertainty in Measurement of Piezoresistive Microsensors.* Dresden: TUDpress (2005).

[148] Schulz, V.; Gerlach, G.; Günther, M.; Magda, J. J.; Solzbacher, F.: Piezoresistive pH microsensors based on stimuli-sensitive polyelectrolyte hydrogels. *Technisches Messen* 77 (2010), 3, 179–186.

[149] Hirokawa, Y.; Tanaka, T.: Volume phase transition in a nonionic gel. *Journal of Chemical Physics* 81 (1984), 12, 6379–6380.

[150] Beltran, S.; Hooper, H. H.; Blanch, H. W.; Prausnitz, J. M.: Swelling equilibria for ionized temperature-sensitive gels in water and in aqueous salt solutions. *Journal of Chemical Physics* 92 (1990), 3, 2061–2066.

[151] Salgado-Rodríguez, R.; Licea-Claveríe, A.; Arndt, K.-F.: Random copolymers of N–isopropylacrylamide and methacrylic acid monomers with hydrophobic spacers: pH–tunable temperature sensitive materials. *European Polymer Journal* 40 (2004), 8, 1931–1946.

[152] Knörgen, M.; Arndt, K.-F.; Richter, S.; Kuckling, D.; Schneider, H.: Investigation of swelling and diffusion in polymers by ^{1}H NMR imaging: LCP networks and hydrogels. *Journal of Molecular Structure* 554 (2000), 1, 69–79.

[153] Ferse, B.: *Photopolymerisation–Grundlagen neuer Synthesen für mechanisch stabile und temperatursensitive Nanokomposithydrogele.* Ph.D. thesis, TU Dresden (2010).

[154] GONG, J.; KATSUYAMA, Y.; KUROKAWA, T.; OSADA, Y.: Double-network hydrogels with extremely high mechanical strength. *Advanced Materials* 15 (2003), 14, 1155–1158.

[155] OKUMURA, Y.; ITO, K.: The polyrotaxane gel: A topological gel by figure-of-eight cross-links. *Advanced Materials* 13 (2001), 7, 485–487.

[156] HARAGUCHI, K.; TAKEHISA, T.: Nanocomposite hydrogels: A unique organic-inorganic network structure with extraordinary mechanical, optical, and swelling/de-swelling properties. *Advanced Materials* 14 (2002), 16, 1120–1124.

[157] FERSE, B.; RICHTER, S.; ECKERT, F.; KULKARNI, A.; PAPADAKIS, C. M.; ARNDT, K.-F.: Gelation mechanism of poly(N-isopropylacrylamide)-clay nanocomposite hydrogels synthesized by photopolymerization. *Langmuir* 24 (2008), 21, 12627–12635.

[158] FERSE, B.; GRAF, M.; KRAHL, F.; ARNDT, K.-F.: Photopolymerization as alternative concept for synthesis of poly(N-isopropylacrylamide)-clay nanocomposite hydrogels. *Macromolecular Symposia* 306-307 (2011), 1, 59–66.

[159] MÜLLER, U.: *Inorganic Structural Chemistry*. Inorganic Chemistry, Chichester: John Wiley & Sons (2006).

[160] FERSE, B.; RICHTER, S.; ARNDT, K.-F.; RICHTER, A.: Investigation of gelling aqueous clay dispersions with dynamic light scattering. *Macromolecular Symposia* 254 (2007), 1, 378–385.

[161] MAGONOV, S.; ELINGS, V.; WHANGBO, M.-H.: Phase imaging and stiffness in tapping-mode atomic force microscopy. *Surface Science* 375 (1997), 2-3, L385–L391.

[162] KENDALL, K.; TABOR, D.: An ultrasonic study of the area of contact between stationary and sliding surfaces. *Proceedings of the Royal Society of London. A. Mathematical and Physical Sciences* 323 (1971), 1554, 321–340.

[163] HARAGUCHI, K.; TAKEHISA, T.; FAN, S.: Effects of clay content on the properties of nanocomposite hydrogels composed of poly(N-isopropylacrylamide) and clay. *Macromolecules* 35 (2002), 27, 10162–10171.

[164] NAFICY, S.: *Conductive tough hydrogels*. Ph.D. thesis, University of Wollongong (2011).

[165] TRANCHIDA, D.; PICCAROLO, S.; SOLIMAN, M.: Nanoscale mechanical characterization of polymers by AFM nanoindentations: Critical approach to the elastic characterization. *Macromolecules* 39 (2006), 13, 4547–4556.

[166] RADMACHER, M.; FRITZ, M.; HANSMA, P. K.: Imaging soft samples with the atomic force microscope: Gelatin in water and propanol. *Biophysical Journal* 69 (1995), 264–270.

[167] REN, H.-Y.; ZHU, M.; HARAGUCHI, K.: Characteristic swelling-deswelling of polymer/clay nanocomposite gels. *Macromolecules* 44 (2011), 21, 8516–8526.

[168] NIE, J.; DU, B.; OPPERMANN, W.: Swelling, elasticity, and spatial inhomogeneity of poly(N-isopropylacrylamide)/clay nanocomposite hydrogels. *Macromolecules* 38 (2005), 13, 5729–5736.

[169] MUKAE, K.; SAKURAI, M.; SAWAMURA, S.; MAKINO, K.; KIM, S. W.; UEDA, I.; SHIRAHAMA, K.: Swelling of poly(N-isopropylacrylamide) gels in water-alcohol (C1-C4) mixed solvents. *Journal of Physical Chemistry* 97 (1993), 3, 737–741.

[170] ZHANG, Y.; FURYK, S.; BERGBREITER, D. E.; CREMER, P. S.: Specific ion effects on the water solubility of macromolecules: PNIPAM and the Hofmeister series. *Journal of the American Chemical Society* 127 (2005), 41, 14505–14510.

[171] ZHANG, Y.; CREMER, P. S.: Interactions between macromolecules and ions: The Hofmeister series. *Current Opinion in Chemical Biology* 10 (2006), 6, 658–663.

[172] CHO, Y.; ZHANG, Y.; CHRISTENSEN, T.; SAGLE, L. B.; CHILKOTI, A.; CREMER, P. S.: Effects of Hofmeister anions on the phase transition temperature of elastin-like polypeptides. *Journal of Physical Chemistry B* 112 (2008), 44, 13765–13771.

[173] ALGAER, E. A.; VAN DER VEGT, N. F. A.: Hofmeister ion interactions with model amide compounds. *Journal of Physical Chemistry B* 115 (2011), 46, 13781–13787.

[174] SCHULZ, V.; GERLACH, G.; RÖBENACK, K.: Compensation method in sensor technology: a system-based description. *Journal of Sensors and Sensor Systems* 1 (2012), 5–27.

[175] DYER, S. A.: *Survey of Instrumentation and Measurement.* Wiley (2001).

[176] KRAUSE, L.: *Dynamische Wägetechnik.* Wissenschaftlicher Verlag Berlin (2005).

[177] BHUSHAN, B. (Ed.): *Nanotribology and Nanomechanics: An Introduction.* Springer (2005).

[178] Bosch automotive aftermarket 2009/2010–oxygen sensors. *Tech. rep.*, Robert Bosch GmbH (2010).

[179] FINGERSON, L. M.; FREYMUTH, P.: *Fluid Mechanics Measurements*, chap. 3. 2nd edn., Taylor & Francis (1996) .

[180] TAVOULARIS, S.: *Measurement in Fluid Mechanics.* Cambridge University Press (2005).

[181] CHE, W. S.; OH, J. H.: Development of force-balance accelerometer with high accuracy for precision motion measurement. *Measurement Science and Technology* 7 (1996), 1001–1011.

[182] STUART-WATSON, D.; TAPSON, J.: Simple force balance accelerometer/seismometer based on a tuning fork displacement sensor. *Review of Scientific Instruments* 75 (2004), 3045–3049.

[183] FORTINA, J.; MARTE, W.; GRÜLLENBERGER, R.; HACKER, A.; HABENBACHER, W.; HELLER, A.; WAGNER, C.; WACHA, P.; SKRABAL, F.: Continuous non-invasive blood pressure monitoring using concentrically interlocking control loops. *Computers in Biology and Medicine* 36 (2006), 941–957.

[184] SHINNERS, S. M.: *Modern Control System Theory and Design.* 2nd edn., John Wiley & Sons (1998).

[185] FRADEN, J.: *Handbook of Modern Sensors: Physics, Designs, and Applications.* 4th edn., Springer (2010).

[186] GAUTSCHI, G.: *Piezoelectric Sensorics: Force, Strain, Pressure, Acceleration and Acoustic Emission Sensors, Materials and Amplifiers.* Springer (2002).

[187] BUDZIER, H.; GERLACH, G.: *Thermal Infrared Sensors: Theory, Optimization and Practice.* Wiley (2010).

[188] DATTA, A.; HO, M.-T.; BHATTACHARYYA, S. P.: *Structure and Synthesis of PID Controllers.* Springer (2000).

[189] VISIOLI, A.: *Practical PID Control.* Springer (2006).

[190] STODOLA, A. B.: Über die Regulierung von Turbinen. *Schweizer Bauzeitschrift* 23 (1894), 17–18.

[191] HURWITZ, A.: Über die Bedingungen, unter welchen eine Gleichung nur Wurzeln mit negativen reellen Teilen besitzt. *Mathematische Annalen* 46 (1895), 273–284.

[192] HOAGG, J. B.; BERNSTEIN, D. S.: Nonminimum-phase zeros - much to do about nothing. *IEEE Control Systems Magazine* 27 (2007), 45–57.

[193] STEWART, J.; DAVISON, D. E.: On overshoot and nonminimum phase zeros. *IEEE Transactions on Automatic Control* 51 (2006), 1378–1382.

[194] NEWTON, G. C.; GOULD, L. A.; KAISER, J. F.: *Analytical Design of Linear Feedback Control.* Wiley (1964).

[195] SCHULZ, V.; EBERT, H.; GERLACH, G.: A closed-loop hydrogel-based chemical sensor. *IEEE Sensors Journal* (2012), accepted for publication.

[196] SCHILP, A.; LÄRMER, F.: *Anisotropic, fluorine-based plasma etching method for silicon.* Patent number WO 1998/037577, Aug. 27, 1998.

[197] MANAVI-TEHRANI, I.; RABIEE, M.; PARVIZ, M.; TAHRIRI, M. R.; FAHIMI, Z.: Preparation, characterization and controlled release investigation of biocompatible pH-sensitive PVA/PAA hydrogels. *Macromolecular Symposia* 296 (2010), 1, 457–465.

[198] JAGUR-GRODZINSKI, J.: Polymeric gels and hydrogels for biomedical and pharmaceutical applications. *Polymers for Advanced Technologies* 21 (2010), 1, 27–47.

[199] RICHTER, A.: *Quellfähige Polymernetzwerke als Aktor-Sensor-Systeme für die Fluidtechnik.* Fortschritt-Berichte: Mess-, Steuerungs- und Regelungstechnik, VDI-Verlag (2002).

[200] ADIGA, S. P.; JIN, C.; CURTISS, L. A.; MONTEIRO-RIVIERE, N. A.; NARAYAN, R. J.: Nanoporous membranes for medical and biological applications. *Wiley Interdisciplinary Reviews: Nanomedicine and Nanobiotechnology* 1 (2009), 5, 568–581.

[201] FURNEAUX, R. C.; RIGBY, W. R.; DAVIDSON, A. P.: The formation of controlled-porosity membranes from anodically oxidized aluminium. *Nature* 337 (1989), 147–149.

[202] MIRZA, A. R.; AYON, A. A.: Silicon wafer bonding: Key to MEMS high-volume manufacturing. *Sensors* 15 (1998), 12, 24–33.

[203] SCHMIDT, M. A.: Wafer-to-wafer bonding for microstructure formation. *Proceedings of the IEEE* 86 (1998), 8, 1575–1585.

[204] SUH, K. Y.; LANGER, R.; LAHANN, J.: Fabrication of elastomeric stamps with polymer-reinforced sidewalls via chemically selective vapor deposition polymerization of poly(p-xylylene). *Applied Physics Letters* 83 (2003), 20, 4250–4252.

[205] SCHNEIDER, F.; FELLNER, T.; WILDE, J.; WALLRABE, U.: Mechanical properties of silicones for MEMS. *Journal of Micromechanics and Microengineering* 18 (2008), 6, 1–9.

[206] LÖTTERS, J. C.; OLTHUIS, W.; VELTINK, P. H.; BERGVELD, P.: Polydimethylsiloxane as an elastic material applied in a capacitive accelerometer. *Journal of Micromechanics and Microengineering* 6 (1996), 1, 52–54.

[207] MCDONALD, J. C.; WHITESIDES, G. M.: Poly(dimethylsiloxane) as a material for fabricating microfluidic devices. *Accounts of Chemical Research* 35 (2002), 7, 491–499.

[208] BHATTACHARYA, S.; DATTA, A.; BERG, J.; GANGOPADHYAY, S.: Studies on surface wettability of poly(dimethyl) siloxane (PDMS) and glass under oxygen-plasma treatment and correlation with bond strength. *Journal of Microelectromechanical Systems* 14 (2005), 3, 590–597.

[209] I/O Module, Datenlogger–USB und Ethernet Mini-Messlabore: Preisgünstiges Minimesslabor LabJack U3, ExaLab U3. Technical report, Meilhaus Electronic (2009), www.meilhaus.com/go/labjack.

[210] RICHTER, A.; PASCHEW, G.; KLATT, S.; LIENIG, J.; ARNDT, K.-F.; ADLER, H.-J. P.: Review on hydrogel-based pH sensors and microsensors. *Sensors* 8 (2008), 1, 561–581.

[211] ORLOWSKI, P. F.: *Praktische Regelungstechnik*. 9th edn., Heidelberg, Berlin: Springer (2011).

[212] LUNZE, J.: *Regelungstechnik 1*. 7th edn., Heidelberg, Berlin: Springer (2008).

[213] HORKAY, F.; HAN, M.-H.; HAN, I. S.; BANG, I.-S.; MAGDA, J. J.: Separation of the effects of pH and polymer concentration on the swelling pressure and elastic modulus of a pH-responsive hydrogel. *Polymer* 47 (2006), 7335–7338.

[214] LJUNG, L.: *System Identification – Theory for the User*. 2nd edn., New York: Prentice Hall PTR (1999).

[215] JUHRIG, A.: *Entwurf, Aufbau und Charakterisierung eines miniaturisierbaren Aktors hoher Energiedichte.* Diploma thesis, TU Dresden (2012).

[216] GANTMACHER, F. R.: *Theory of Matrices*, vol. II. New York: Chelsea (1959).

List of relevant publications

Journal articles

1. V. Schulz, H. Ebert, G. Gerlach: A closed-loop hydrogel-based chemical sensor. *IEEE Sensors Journal* 13 (**2013**), 994-1002.
2. V.Schulz, G. Gerlach, K. Röbenack: Compensation method in sensor technology: a system-based description. *Journal of Sensors and Sensor Systems* 1 (**2012**), 5–27.
3. V. Schulz, S. Zschoche, H. P. Zhang, B. Voit, G. Gerlach: Macroporous smart hydrogels for fast-responsive piezoresistive chemical microsensors. *Procedia Engineering* 25 (**2011**), 1141–1144.
4. V. Schulz, G. Gerlach, M. Günther, J. J. Magda, F. Solzbacher: Piezoresistive pH microsensors based on stimuli-sensitive polyelectrolyte hydrogels. *Technisches Messen* 77 (**2010**), 179–186.
5. J. Sorber, G. Steiner, V. Schulz, M. Günther, G. Gerlach, R. Salzer, K.-F. Arndt: Hydrogel-based piezoresistive pH sensors: Investigations using FT-IR attenuated total reflection spectroscopic imaging. *Analytical Chemistry* 80 (**2008**), 2957–2962.

Conference contributions

6. V. Schulz, B. Ferse, A. Große, K.-F. Arndt, G. Gerlach: Clay nanocomposite hydrogels applied to MEMS-based chemical microsensors. *Proceedings of IEEE NEMS Conference* **2012**, 103–104.
7. V. Schulz, H. Ebert, G. Gerlach: A novel closed-loop hydrogel-based pH sensor. *Proceedings of IEEE Sensors* **2011**, 592–595.
8. V. Schulz, B. Ferse, K.-F. Arndt, G. Gerlach: Characterization of organic-inorganic clay nanocomposite hydrogels and integration as smart building blocks into chemical microsensors. *2nd International Conference on Multifunctional, Hybrid and Nanomaterials*, Strasbourg, France, March 6–10, **2011**, Book of Abstracts, B24.

9. M. Günther, V. Schulz, G. Gerlach, T. Wallmersperger, F. Solzbacher, J. J. Magda, P. Tathireddy, G. Lin, M. P. Orthner: Hydrogel-based piezoresistive biochemical microsensors. In: Y. Bar-Cohen (Hrsg.): Electroactive Polymer Actuators and Devices (EAPAD) **2010**, *Proceedings of SPIE*, Vol. 7642, 764227.

10. V. Schulz, M. Günther, G. Gerlach, J. J. Magda, P. Tathireddy, L. Rieth, F. Solzbacher: In vitro investigations of a pH- and ionic-strength-responsive polyelectrolytic hydrogel using a piezoresistive microsensor. In: Y. Bar-Cohen (Hrsg.): Electroactive Polymer Actuators and Devices (EAPAD) **2009**, *Proceedings of SPIE*, Vol. 7287, 728712.

About the author

	Personal data
Name	Volker Schulz
Date of birth	19.02.1982
Born in	Bad Salzungen
	Education
1988–1998	Realschule, Bad Salzungen (mittlere Reife)
1998–2001	Gymnasium, Eisenach (Abitur)
	Academic background
05/2008–07/2013	Research associate/Ph.D. student, Solid-State Electronics Laboratory, TU Dresden
07/2010–10/2010	Visiting researcher, Nano-/Micro-System Laboratory, Kyoto University (Japan)
28.01.2008	Diploma (Dipl.-Ing. degree)
10/2002–01/2008	Study of electrical engineering, TU Dresden Focus: microsystems, semiconductor and sensor technology
11/2006–04/2007	Internship, IMEC, Leuven (Belgium)
	Awards
06/2013	Eugen-Hartmann-Preis 2013, awarded by the VDI/VDE-Gesellschaft für Mess- und Automatisierungstechnik (GMA)
07/2009	Professor-Schwabe-Preis 2009 (for Diploma thesis), awarded by the Professor-Schwabe-Stiftung